Job studies and industrial relations

Job studies and industrial relations

Ideas about efficiency and relations between the parties of the labour market in Sweden 1920–1950

by HANS DE GEER

Almqvist & Wiksell International
Stockholm – Sweden

ISBN 91-22-00479-3

Translated from the Swedish by William Barrett, Stockholm

Published with support from
Riksbankens jubileumsfond (The Bank of Sweden Tercentenary Fund)
and Humanistisk-samhällsvetenskapliga forskningsrådet
(Swedish Council for research in the Humanities and Social Sciences)

Graphic design by Jerk-Olof Werkmäster
Printed in Sweden by Almqvist & Wiksell Tryckeri AB, Uppsala 1982

Contents

Acknowledgements

This investigation was carried out mainly during the period 1977–1979 and the final Swedish manuscript was completed in the summer of 1980. The study started within the framework of the research project Kris och Krispolitik (Crisis and Crisis Policy in Scandinavia during the Interwar Era) and links up with my earlier research Rationaliseringsrörelsen i Sverige—Effektivitetsidéer och socialt ansvar under mellankrigstiden (The Rationalization Movement in Sweden). This carried over to and was completed within the project TES, Teknologi och effektivitetsidéer i samhällsomvandlingen 1920–1970 (Technology and ideas of efficiency in the social transformation 1920–1970), in which was studied the spread and breakthrough of technology and ideas with an economic tinge on other sectors of society.

It is not at all unusual that research seems to be a lonely job. But really there is an intricate social relationship lying behind the results for which the researcher is solely responsible. Various contributions from many quarters have constituted the necessary conditions.

Financial support has been provided by Riksbankens Jubileumsfond (The Bank of Sweden Tercentenary Fund), Humanistisk-Samhällvetenskapliga Forskningsrådet (The Swedish Council for Research in the Humanities and Social Sciences) and Forskningsrådsnämnden (Swedish Council for Planning and Coordination of Research). The first two mentioned have also met the costs of translating and publishing the work.

The basic material for the presentation, besides a number of earlier researchers and authors, has essentially been found at the organizations investigated: Sveriges Verkstadsförening (The Swedish Metal Trades Employers Association), Svenska Metallindustriarbetarförbundet (The Swedish Metal Workers Union), Sveriges Textilindustriförbund (The Swedish Textile Employers Association), Svenska Beklädnadsarbetareförbundet (The Swedish Textile Workers Union) and Landsorganisationen i Sverige (The Swedish Trade Unions Confederation). Of those who have kindly helped me I should specially like to mention chief archivists Gunnar Åström, Sveriges Verkstadsförening, and Ragnhild Lindberg, Svenska Metallindustriarbetarförbundet, Svante af Winklerfelt, director of Sveriges Textilindustriförbund, and the former secretary of Svenska Beklädnasarbetareförbundet, Georg Lindgren.

Important parts of the source material have been found in the Swedish National Archives and the great help given by the personnel there is gladly acknowledged.

My working environment has been, as always, happily provided at the Historical Institute of the University of Stockholm and at the university branch library at 59 Odengatan, Stockholm, where the staff produced the necessary literature most willingly.

Good fellowship and intellectual companionship has been provided by friends and colleagues at the Institute, and in this context I should especially like to mention my friends in the project TES, professor Gunnar T Westin and docent Per Thullberg. Professor Per Erik Back has kindly read parts of the earlier version of the manuscript.

Kerstin Israelsson, secretary to the research project has, in this as in the other work covered by the project, made an invaluable contribution. Mr William Barrett has energetically and with great interest rendered the manuscrip into English.

For all of these contributions I am deeply grateful.

But research work also gives a distinctive character to one's private life. My nearest and dearest, to whom this book is dedicated, have had to share the situation, both its positive and negative sides. I thank them warmly for their help and understanding.

Stockholm i February 1982
Hans De Geer

PART 1

Points of departure

CHAPTER 1

Problems and sources

1. Introduction

This investigation deals with the question of how the technology of modern production influenced the relationships between the parties of the labour market during the period from 1920 to 1950. The term 'production technology' will be used to cover the broad reality and include both technical as well as organizational elements. In view of the fact that the investigation consists of parallel studies of two main industrial areas, essentially organizational factors in the new technology will receive the limelight. For their range extends to many areas of industrial production.

In an earlier study I have investigated the ideas in the Swedish rationalization movement and their extension over Swedish industrial life during the interwar years.[1] Here were found especially two characteristics of these ideas about efficiency during this breakthrough period. These were 'costs per unit', that is production costs, and 'time and motion study', or 'job study'.

It was decided to follow time and motion studies and their influence on the relationships between the opposing parties parties of the labour market. The term 'time and motion studies' here incorporates both time studies and method studies.

Time and motion studies may generally be said to have had two functions. They could be a lead in the development of methods and processes within industry by revealing a lack of balance in the interplay between the production factors utilized or by showing other sources of loss. They could also be used for guidance in a pay system based on performance, where the earnings (usually piece wages) were calculated proceeding from the values of the norms arrived at through time and motion studies or from such studies carried out in every special case.

In both of these functions—as a factor in a more general rationalization process and as a part of the pay system—time and motion studies were to influence the relationships between the parties of the labour market. It was initially as a basis for the setting of piece wage rates that time and motion studies were discussed at the collective negotiations on the national agreements. The alterations over time carried out there are those to be more closely analyzed in what follows.

The development in two different industrial areas covered by agreements will be investigated and used as the starting point of a more general discussion about the consequences of the new technology. One of these is the metal trades industry, which was the pioneering sector of the rationalization movement, where basically the agreement between the Metal Trades Employers Association and the Metal Workers Union regulates the conditions. The other area also affects an industrial branch, the textile industry, which was early made to feel the effect of rationalization and time and motion studies, where the contracting parties were the Textile Employers Association and the Textile Workers Union. There are similarities and marked differences between the metal trades and the textile industry which, by means of comparisons in the following, can be taken as premisses for the line of reasoning used.

In each of these agreement areas will be discussed how the agreement was altered so as to give consideration to and regulate the new reality taught by time and motion studies. Attention here will be focussed on the changes in the parts of the national agreements that regulated piece wages. It seems, however, to be less useful here to limit too strictly the area of examination to only those paragraphs directly concerned with time and motion studies. Alterations in the texts of the agreements did not occur in isolation and therefore should not be so considered. Therefore usually it will be the case that the whole of the piece wage paragraph is considered, and often also other parts of the national agreement.

Not only the actual development of the agreements will be shown and analyzed but also the parties' actions at the central negotiations. This means that those proposals and opinions that did not lead to any results will also be considered. In addition a special point will be made of the internal reactions of the organizations affected by the time and motion study questions, often in conjunction with, or in some other way linked to, the preparations for or the carrying out of the collective negotiations.

The present investigation is based on source material from the central organizations and thus considers the various problems from a central angle of approach. It is important to note this, especially in such a context. Time and motion studies and other rationalization measures often appear to be quite different, depending on the point of view obtaining. The ability of human beings to isolate the rational aspects of a question has a tendency to increase the less they themselves are directly affected by the negative consequences of the rational actions.[2] The central outlook, which through the selection of sources dominates this presentation, captures the local opinions only to the degree that these were advanced to the central organs and congresses, to policy conferences about the agreement, or otherwise. Thus the investigation is based on material from the general councils and executive boards of unions or/employers, and on the collective negotiations for the national agreements, which has kindly been placed at my disposal by all of the main organizations concerned. This has made it possible to achieve an important quality in the investigation: to be able so far as possible to cover both the employers' and the workers' sides in the parties' relations. Unfortunately it has not been possible to maintain this reciprocity at the central organizational level. The Trade Union Confederation's archives have been opened to me, but the Employers Confederation, which seems remarkable for those engaged in a serious research, has not considered itself able to accede unreservedly to the clearly detailed wishes concerning access to material that could have thrown some light on the growth of the time and motion study agreement of 1948. (Regarding sources see further p. 22.)

Earlier research about time and motion studies or closely related problems is scarce and has usually had a somewhat different direction than in the present investigation.[3] Such matters have usually been the subject of attention from sociological quarters and the consequences of increased efficiency on actual working conditions (redesign of job, division of labour) have often come to the fore. However, in this investigation it is rationalization, the role of the new technology in the relationships between the parties, on which attention is focussed. This also brings with it a comparatively greater interest for the influence time and motion studies had on questions of pay.

2. Balance of relations between the parties

During the period covered by this investigation, the Swedish labour market was characterized by acknowledged industrial relations. The metal trades agreement of 1905 and the implications of the so-called "December compromise" stand out as the basis of these. This latter agreement, concluded in December 1906 between The Swedish Employers Confederation and the Swedish Trade Union Confederation meant that the workers' right to organize was conceded as well as the manner in which their organizations carried on negotiations and concluded agreements concerning questions about the allocation of financial results, with special reference to the setting of wages. In return the workers admitted the employers' right to direct and allocate work and also to engage and discharge personnel. This right to decide how production should be planned was to be found from then on in the Employers Confederation's standing orders under varying paragraph numbers including 23 and 32, and it was presumed that the employers' associations when becoming members of the confederation inserted a similar paragraph in the collective agreements they made. The Metal Trades

Employers Association, however, regarded this principle as so natural and so basic that it was never written into the metal trades agreements. The right of the employers' side was also tacitly acknowledged by its counterparts, so that when this association joined the Employers Confederation in 1917 an exception was made: to insert a "paragraph 23" would change nothing in reality but cause a lot of irritation to the union side.

On the basis of the December compromise a balance developed between the recognized parties of the labour market. It was not that this compromise created a balanced position. It is perhaps more accurate to say that the compromise set limits to how far the balance could be swung to the advantage of one or other of the parties. The power of the employers could not be so marked that it affected the rights of association and negotiation; the influence of the workers' organizations might not impinge upon questions relating to the organizing of work.

Undoubtedly during the seventy years that have followed the December compromise changes have occurred in these limits to the scope of the parties' negotiations. Already by the middle of the 1930s the workers had gained a certain influence on questions concerning engagement and discharge. During the 1970s the Codetermination Act 1977 has been directed against the remainder of the employers' exclusive right to decide on matters of production: "the abolition of paragraph 32".

For the *employers*—at least for the organizations—it has seemed vital to defend management's exclusive sphere, to defy all attempts by the workers to gain influence and joint determination on production questions. This attitude has also marked the employers' actions *within* the framework that limited the negotiable area. In this investigation one policy for their actions is considered to have been to ensure initially the *employer's* (the management's) *freedom of manoeuvre*. This can be called a fundamental hypothesis about the aim of the employers' aspirations. It implies an aspiration to *limit to a minimum the issues* which were regulated in agreements. The areas that were not subject to regulation by agreement were left for independent decisions or to wholly local agreements, where the locally stronger part had the upper hand. This endeavour to ensure freedom of manoeuvre further led, in the cases where regulation by agreement could not be avoided, to attempts to mould the *central agreement provisions* so that they *de facto* guaranteed local freedom of manoeuvre.[4] This does not exclude a strategy on the organizational level that aimed at centralization and in this way the establishment of strong contracting organizations. Thereby could best be advanced the stated aims of the national agreements' objective content.

Piece wage questions were of course those about the allocation of financial results, but the regulation of these by agreement followed the pattern mentioned. For the employers the right to independent piece wage agreements was a strategically important bastion which had to be defended to the bitter end. The provisions in the national agreement, in the period covered by the beginning of the investigation, were therefore rather few, but increased with every new negotiation of the agreement. On the employers' part, it is presumed, an attempt was made insofar as it was possible to avoid the regulation by agreement of piece wages or at least to see that the provisions were so designed that the greatest possible freedom of action was obtained at the local level.

It should be noted that the attempts of the employers to limit the agreement's contents to a minimum may not be taken to apply to all areas concerned. For example the reverse applied to the common rules of conduct on which the parties agreed.

For the *workers* the diametrically opposite aspirations seemed natural. The desire to alter the limits of areas of negotiation has been mentioned as well as the obvious result of this. Nevertheless it has been remarkable for the Swedish trades union movement that it has accepted the stated limitation of areas of activity to such a high degree. Sociologists are of the opinion that Swedish workers have had an "instrumental" view of work.[5] It has been regarded solely as a tool, an "instrument" with which to create the material conditions for a richer life in leisure time.

This instrumental attitude has also to a large degree influenced the aims of the organizations. There has been a tendency initially to make sure the pay conditions were right and secondly to consider such working conditions as only indirectly influenced questions of pay. This aim can be explained by reference to a number of factors. The idea of the December compromise was to give the organizations precisely this role. The desire to eradicate what remained of a more patriarchal employer relationship also led to an emphasis of the pecuniary compensation.

The tendency to proceed with the instrumental role's limited field of ambition was a restraining factor for the trade union organizations' actions in all other questions than those that unmistakeably dealt with the allocation of financial results. But on the other hand this concentration could also increase the pressure behind the demands raised by these organizations about pay.

The basic aspiration of the trade unions was to change the real balance of influence and power. This should be achieved by a mobilization of the masses and through a channelling of their weight into organizational forms. The consequence of this way of looking at things is that the greater the masses mobilized, the greater the weight behind the organizations' actions. This means that the demand for results from activities carries with it a tendency to centralization. It is at a central level that the organizations' authority against the employers is greatest.

These circumstances lead to the following hypotheses about the workers' general desires in collective negotations with the opposite party. Here it is presumed initially that the workers' aspirations were aimed at creating *central, unconditional provisions,* for the matters concerned. Secondly, the workers' organization had the idea of obtaining agreements with provisions which implied that *central guarantees* were created so that the workers were *ensured central or local influence* in the decision process. Thus the employers' local superiority should not play such a great part.

Expressed in terms of the piece wage problems' realities, these aspirations should be expressed in the following manner. Initially the workers' side should strive for central, binding provisions for, as examples, which jobs should be done at piece wages, how high the minimum earnings should be, and also in which way and under what conditions the piece wage should be determined or changed.

Secondly, the workers should try to obtain guarantees that the local workers, in negotiations with managements, or the trade union, obtained such knowledge and such influence that they could negotiate on the same footing.

3. Time and motion studies in the metal trades and textile industries

Given this balance and given the parties' common efforts, the question may be asked as to what effects could time and motion studies produce.

Time and motion studies were a tool used to make industry more efficient. They contributed to changes in production methods and in the way of calculating earnings. In the present investigation, time and motion studies are regarded as *a new technique,* one result of an altered technological production.[6] It is therefore necessary that this investigation shall also throw some light on the more general problems concerning *the part played by changed technology in the relationships between the parties of the labour market.*

In a recently published presentation, Walter Korpi discusses which factors influence job design. He discerns two schools among earlier Swedish researchers. One school derives working conditions directly from technological production-conditions, from the company's environment and size. Against these Korpi stresses the idea held by the other line of thought, that the organization of work is a question of power. The group with the power of decision at the workplace arranges the conditions so that its own position of strength is stressed. This can occur for example in the allocation of the work: planning and performance are differentiated.[7]

In a similarly newly-published essay, Bertil Gardell presents a closely related line of thought. He discusses the attitudes of different social groups to technology and maintains that these were not—as many of the zealous champions of progress thought in the 1930s—mainly dependent on the degree of information and explanation. Instead the way in which technique should be regarded was dependent on the varying possibilities of influence and of the actual material conditions of interplay between man and technology.[8]

The presentations cited are of some help in formulating two pairs of hypotheses on the parties' relation to technique/time and motion studies. The first pair may be expressed in the following way: (1) *The party introducing time and motion study techniques does it for the purpose of strengthening his position in the relationship between the parties.* (2) *The parties' attitudes to time and motion studying techniques are directly dependent upon their possibilities of influencing the utilization.* By "attitudes' here is meant points of view expressed not only in words but also in action.

(It should be noted that this applies mainly to the use of time and motion studies in the setting of piece wages—in the relationship between the parties in a limited sense—whilst time and motion studies used as method studies may have the same effects but in a more indirect way.)

An alternative pair of hypotheses develops the idea of rationalization as an "objective" process: (1) Job studies are introduced (requested) by either party because they constitute *the best way* (the most efficient, the most just, etc.) of evaluating work in relation to the norms for earnings and the designing of the work process. Such an exclusion of a tactical organizational approach seems doubtful however in a study of trade union action, but would certainly seem more natural in the investigation of rationalization questions in a private company. (2) The attitude of the workers to time and motion studies depends on the degree of information provided.

To express this in a more applied manner, the first pair of hypotheses may be said to imply the following presumptions. The employers started to use time and motion studies as a tool in the setting of piece wages because the studies provided such arguments for the employers' cause that for various reasons were difficult for the workers to meet. One reason for the difficulties could of course be that with the assistance of time and motion studies it was actually possible to obtain a more indisputable basis for the setting of piece wages. Another reason could be that the technical complexity of time and motion studies was a closed book for the average worker in situations where he had to make an agreement on a piece wage.

The workers were also negative to time and motions studies to the same degree that they had no chance of insight into, and influence over, them. Time and motion studies in the hands of the employers seemed to present a threat to the relations of the parties of the labour market. The threat could be met either by using the same weapon or by forbidding or depying its use.

It is quite easy to see that had these different attitudes been urged consistently they would have set the organizations on a collision course. But the collision did not occur. This fact calls attention to the second pair of hypotheses. The central organizations gradually reached remarkable unity with regard to rationalization. In 1948 the Trade Union Confederation and the Employers Confederation entered into a collaboration agreement on time and motion studies in industry. This meant that one or the other or both of the parties concerned must have waived their demands or at least given some kind of consent. One of the central tasks of this investigation is to show how the parties were compelled to these compromises in basic aspirations, and further to confirm when this occurred, under what conditions and to whose advantage.

But in testing the hypotheses it must be borne in mind that the level of investigation influences the results and the possibilities of generalization from these. It is, as has been shown earlier, neither on the level of the central confederations nor on that of a single workshop that the investigation results are to be found, but at the level of the trade unions' organizations. Parallel studies of two agreement areas should make possible a comparison, which may reveal explanations that would have been more difficult to discover in a study of a single branch.

The metal trades industry was the pioneer sector for the whole of the industrial world in attempts at rationalization. Time and motion studies started in Swedish workshops even before the World War I. The irritation about time and motion studies during the 1940s was greatest in the metal trades sector.

The textile sector also lay well to the fore with an intensification of rationalization and time and motion studies, especially from the 1930s. However, even a very brief glance shows that time and motion studies in the textile industry were not at all the practical, political and union problems that they were in the metal trades workshops. This striking difference calls for a discussion of the explanatory factors.

To constitute a basis for the assessment of the different roles time and motion studies played in both branches certain questions can be formulated. These questions concern different factors: the actual utilization of time and motion studies, the character of the production or characteristics of the pay system, the agreements or the organizations.

First of all, can the quieter development of the textile agreement's area be explained simply by the fact that the time and motion studies were used to a lesser degree than within the metal trades industry? Did the method never really make a breakthrough on such a broad front in the textile industry? Was it the case that time and motion study methods were transferred from the metal trades area only after a number of "growing pains" had been eradicated? Had the attempts at rationalization actually been pushed harder in the metal trades industry than in the textile industry? What part was played by changes in trade?

Perhaps the explanation of why it was possible to avoid a high degree of irritation within the textile industry is to be found in the conditions of production? Of what importance was the industrial structure? Was the production process in the textile industry more controlled by the machines' capacity and by the quality of the raw material so that the individual worker's contribution was unable to influence the volume of production and the piece wage earnings immediately? The less directly time and motion studies were able to influence earnings perhaps the less was the irritation? Was it really that textile production was so much more uniform, that the product series were so much longer, that the need to set new piece wages was so much less than in the metal trades workshop, where the mixed production continually compelled the setting of new wages?

Or perhaps the pay system was different, thus causing disparities? Was the volume of piece wages considerably lower in the textile industry?

Was the construction of the agreements different? Were possibly the provisions of the textile agreement better adjusted than in the metal workers agreement for the purpose of solving the problems that time and motion studies presented to the parties?

Perhaps there was a difference in the organizations' ways of dealing with the problems? Was the Textile Employers Association more accommodating than the Metal Trades Employers' Association? Was the Metal Workers Union more militant than its colleagues in the textile industry? If this was the case what was the reason: a greater measure of political radicalism in the metal workers, a larger proportion of women and underage workers in the textile branch, or some other factor? The detailed presentation of the development in both of these areas of agreement will provide the foundation for a discussion about these problems in the final part of this work.

4. Lawyers and engineers

A further important characteristic in the development of the society of modern production technology will also be considered: the question of the vertical grade structure of work and professionalization, which has made itself felt even in the area of trade union relations. It is the intervention of experts in negotiation situations, it is the *engineers*' and *lawyers*' importance in the construction and understanding of agreements where production technology became increasingly complicated, when the negotiation system laboriously sought to draw up agreements with reference to the numerous facets presented by reality and where labour legislation gradually attempted to balance reality with intention.

In their studies of the Labour Court, Lennart Geijer and Folke Schmidt, with their legal backgrounds, have enquired how the representatives of the parties of the labour market functioned in the court both as interpreters and judges.[9] Investigations of the metal trades and textile agreements' development make it possible here to present in a special sense a contrary view of the Labour Court's function. Here it is not the behaviour of the parties of the labour market that comes to the fore but the court's way of guiding, of setting up norms and leading the development in the areas of agreement that formed the main interest.

It is then demonstrated that there exists an amazing parallelism between the basic evaluations on which the collective agreement legislation was constructed and the Labour Court followed, and those that formed the basis for the use of time and motion studies. In both cases the line of argument proceeded from the assumption that conflicts on the labour market could be divided up into "objective" requirements on the one hand and real conflicts of interest on the other. The purpose of time and motion studies was to lay down with greater accuracy the "objective" foundation for the setting of piece wages and, parallel with this, it was thought that indisputably the Labour Court could arrange the legal conditions correctly and make decisions where interpretation was required. This basic parallelism leads to the following hypothesis for the court's actions: that the Labour Court in appropriate cases tended to develop a trend towards an increased use of such techniques that claimed to increase the "objective" content of the problems in dispute. *The Labour Court would thus through its interpretations of the current agreements encourage the tendency to utilize methods like time and motion studies in industry.*

The other group of experts was the engineers. The ambitions to be found among engineers at the beginning of the interwar years to form a mediation group between representatives of labour and capital I have referred to in another context.[10] The whole of the rationalization movement, and particularly time and motion studies, was a manifestation of this engineers' ideology. Thus to accept time and motion study therefore meant, to a certain degree, to encourage this engineer-mentality (it can just as well be called rationalism) which interfered in the relationship between the parties.

Linked to this mentality, the tendency to pursue logical arguments from the stated conditions with the greatest consistency, were usually found ideological characteristics that provided measurable technical and economic basic values but which failed to identify other not easily quantified human qualities.

Either the engineers obtained influence over the parties or the representatives for the parties appropriated the confidence of engineers. An important task in the present investigation is to try to decide when and how the skill of engineers became unavoidable for the parties and therefore impossible to handle for those who were unable to make the technical calculations.

5. Source material of the investigations

As was mentioned earlier, this investigation is based mainly on unpublished source material from the organizations concerned.

The source material preserved by the *Metal Trades Employers Association* (Verkstadsföreningen) is abundant and the association's archives are rather well-arranged. The minutes of the general council have been made available as well as files with documents preserved from negotiations concerning national agreements. I have also had access to the proceedings of internal committees, for example the time and motion study committee. It is easy to locate items in the association's archives, even for researchers, because of the extremely helpful registration.

The Metal Trades Employers Association has earlier been made the subject of a history by Georg Styrman, who was its managing director for many years. It is of course based on internal material and personal experiences but is in its character more of a party contribution than an attempt at writing objective history.

The archives of the opposing side, the *Swedish Metal Workers Union* (Svenska Metallindustriarbetareförbundet), have also been made available. In addition to the printed series of minutes from congresses and policy conferences on the agreements, the unpublished minutes of the general council and the executive board have been utilized, together with the documents attached to them. As was the case with the Metal Trades Employers Association, the material supplied has provided a very useful insight into the executive board's internal discussions. The policy conferences reflect the growth of opinion at the branch level.

In the case of the Metal Workers Union there is a four-volume history of the union's activities from its foundation up till 1956. The sections dealing with the years after 1925 are written by Professor Per Erik Back who, unlike Georg Styrman, had definite scientific ambitions. Often in the present text it has been possible to discover a sound basis from Back's presentation.

With regard to the area covered by the *textile agreement* the conditions provided by the available material have not been so helpful as was the case in the metal trades industry. The archive material preserved has been made available to researchers by both parties with great generosity but here, for various reasons, the picture is still somewhat more fragmentary.

The archives of the *Textile Employers Association* contain lucid series of minutes from the general council, the executive board, the national collective negotiations and negotiations about disputes. The minutes from technical conferences and committees on the other hand show certain gaps, presumably caused by the abstraction of documents for destruction. A similar fate has also befallen the association's correspondence from the period being investigated.

The Textile Employers Association's development during the first 50 years has been depicted with great care by the former head of section at the National Social Welfare Board, Nils Bergsten, in a bulky volume marked by its wealth of facts, great accuracy and richness of detail.

The *Swedish Textile Workers Union* combined in 1972 with the neighbouring organizations, the Swedish Clothing Workers Union and the Swedish Shoe and Leather Workers Union. The archives for the years covering the present investigation were then transferred to the popular movements' archives in Norrköping, where the earlier union office had been situated. Minutes from the general council and the representative council and documents from the national collective negotiations have been used for this present study. On the other hand, the Textile Workers Union's organization at that time did not hold regular policy conferences on the agreement at which the branches' demands and wishes were put forward. At certain times such conferences could be called by a decision of the general council. But the more continuous picture of the development of opinion within the union provided by the regular policy conferences of the Metal Workers Union must for the Textile Workers Union be sought in the letters that were reported at the general council's meetings and also attached to its minutes. Certain correspondence has also been preserved in connection with some policy conferences on national collective negotiations.

The Textile Workers Union's development is described in a popular, easily read history, written for the union's 40 year jubilee by the earlier union's representative, also chief editor for the newspaper *Västgötademokraten,* K. J. Olsson. Supplementing this is a volume covering the years 1938–1948 by Bo Tillman, who later became chairman of the union. The history of the Textile Workers Union may in its character be closely compared with Georg Styrman's book about the Metal Trades Employers Association: the jubilee publication, largely devoted to the historic development, has an unmistakably biased perspective.

Where appropriate the material from the organizations' archives has been completed from various sources. Worthy of special mention are the documents left behind by the Conciliation Commissions set up by the government, which have sometimes been utilized. They are stored in the archives of the Public Conciliator's Office at the National Archives under the title of Special Appointments (Särskilda förordnanden). These documents consist of minutes, drafts for tentative feelers, etc., as well as completely unedited, sometimes almost impossible to understand, rapidly scribbled notes. In combination with the material kept by the organizations they have made it possible to illuminate some decisive points during the course of the negotiations analyzed.

PART 2

The Metal Trades Industry

CHAPTER 2

The metal trades industry—a presentation

1. Development of the metal trades industry

The metal trades industry is located in the Swedish official statistics in the subgroups iron and steel manufacture, machine shops, shipbuilding yards, electrical engineering, instruments and clocks, metal products, and gold and silver goods. Combined these amount to about 15 per cent of Swedish industry's production value at the middle of the 1920s. The proportion increased to over 20 per cent during the following decade approaching 25 per cent at the beginning of the 1940s, a figure that following some decline was again reached and exceeded at the end of this decade.

In the metal trades industry were employed 22 per cent of industry's total workforce during 1925, with between 27 and 31 per cent during the 1930s; 33 per cent at the end of the second world war, and 36 per cent in 1949. The figures show that development within the area covered by the metal trades agreement, which will be analysed in the following, affected between a third and one-fifth of Swedish industry and its workers.

In Table 1 is presented statistical information about the different branches of the metal trades industry. It will be seen that the metal trades machine shops were responsible for approximately half of the metal trades production value and that they also employed an equivalent proportion of the total number employed. Comparable from both aspects are iron and steel manufactures, shipuilding and the electrical industry, each with a share of between 10 and 15 per cent of the metal trades production value and employment volume. Instrument and watch and clock factories, other metal manufacturing factories and factories for gold and silver goods were answerable for some few per cent of the whole of the industry.

The development of production value, employment and power consumption in the four larger branches appears from diagrams 1 to 4. The presentation is based in these on index series prepared and proceeding from the 5-year average for the period 1920 to 1924. Production value development is

Table 1. *Metal trades industry in figures*

Source: Official Statistics of Sweden – Industry

	1925	1930	1935	1940	1945	1949
Sales manufacturing value (million Skr)						
Iron and steel	108	127	179	329	432	791
Metal trades machine shops	340	531	668	1 156	1 498	2 975
Shipbuilding	56	105	90	189	270	602
Electrical engineering	92	158	169	316	378	950
Instruments and clocks	11	12	13	22	29	62
Manufactures of metal	37	51	66	98	127	280
Gold and silver goods	12	15	18	16	31	50
Total	656	999	1 203	2 126	2 765	5 710
Number employed (1 000's)						
Iron and steel	16	20	25	31	37	42
Metal trades machine shops	49	66	75	104	135	154
Shipbuilding	11	16	13	20	25	27
Electrical engineering	10	15	17	27	36	46
Instruments and clocks	2	2	2	3	4	5
Manufactures of metal	5	7	9	11	13	16
Gold and silver goods	1	2	2	1	2	2
Total	94	128	143	197	252	292
Production value per manhour (Skr)						
Iron and steel	3.04	2.82	2.91	4.89	6.37	8.69
Metal trades machine shops	3.45	4.07	4.84	6.18	8.35	12.43
Shipbuilding	3.46	4.37	5.04	6.80	11.42	17.41
Electrical engineering	5.61	6.49	6.80	8.42	10.73	15.93
Instruments and clocks	3.45	3.63	3.72	4.76	6.20	7.66
Gold and silver goods	4.59	4.84	5.16	6.93	10.57	13.88
Power installed per employee (h.p.)						
Iron and steel	3.04	2.52	2.23	2.68	2.88	3.77
Metal trades machine shops	2.20	2.04	2.16	2.40	2.54	3.08
Shipbuilding	2.63	2.50	3.27	3.29	3.81	4.75
Electrical engineering	2.20	2.61	2.61	2.66	2.33	2.37
Instruments and clocks	0.80	0.88	0.84	1.23	1.27	1.55
Manufactures of metal	1.26	1.38	1.40	1.80	2.10	2.83
Gold and silverware	0.81	1.09	1.17	1.76	1.35	2.06

stated in current prices: the inflation of the late 1940s has especially influenced the appearance of the curves in a manner that does not clearly reflect changes in the production volume. An obvious influence on the appearance of the production value curves is also caused by the great metal trades conflict in 1945. The decline in that year may be seen as a parenthesis in a more long-term change in the production result.

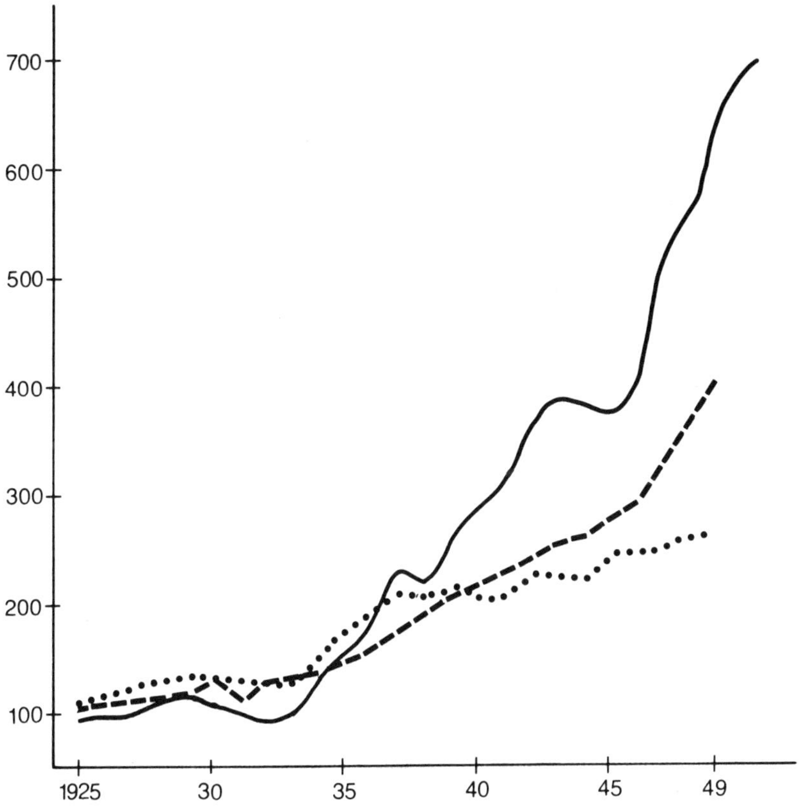

Diagram 1. Iron and steel manufacturing plants.
Production value (—), Number of workers (······) and power installed for direct production (---). Index 1920–1924=100.
Source: Official statistics of Sweden SOS, Industry.

All branches follow the expected trade development: a clear upward tendency during the latter part of the 1920s and the brief decline between 1930 and 1932 when the ascending trend is renewed, in some cases with a break in the curve after the middle of the 1930s, accelerating during the following decade.

The curves for the number of workers show mainly the same profile but are flatter: the workforce is not laid off immediated in bad times but neither is new recruiting initiated rapidly when the economic position improves.

The curves showing motive power installed are characterized by an almost unbroken, even rise. Mechanization in its various phases proceeded almost independently of the development of trade—only the electrical industry here presents a picture of a more unsettled development.

In the comparison between production value and employment is to be found a measure of productivity that is of interest in this investigation.

Diagram 5 shows the development of production value per man-hour in

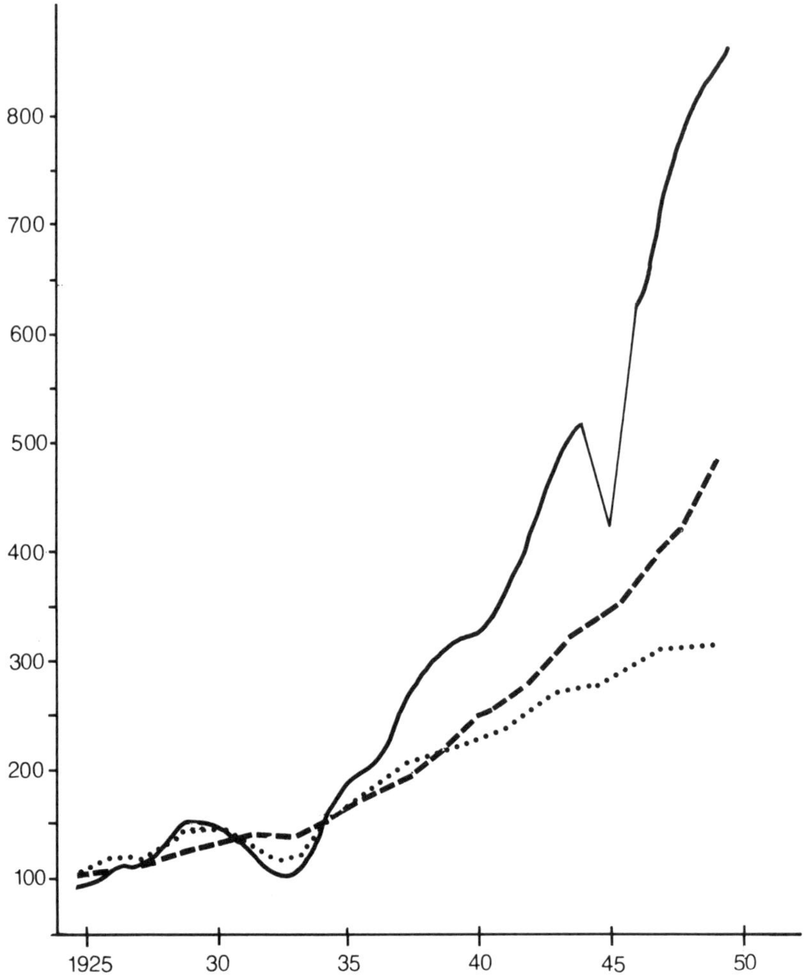

Diagram 2. Metal trades machine shops.
Production value (—), No. of workers (· · ·) and power installed for direct production (---). Index 1920–1924=100.
Source: SOS Industry.

the different branches. As can be seen, in all branches there was a clear increase in productivity, this being most marked the shipbuilding and electrical industries, plainly in the engineering workshops but clearly also in manufactures. But a reservation must also be made here for the influence that price development in the different branches may have had on the calculations. What is clear, however, is that the whole increase in production value can be attributed neither to price development nor to an increased volume of employment. The increase in productivity was a reality.

It is not unnatural to suspect that the increased use of electrical power because of increased mechanization was mainly responsible for the in-

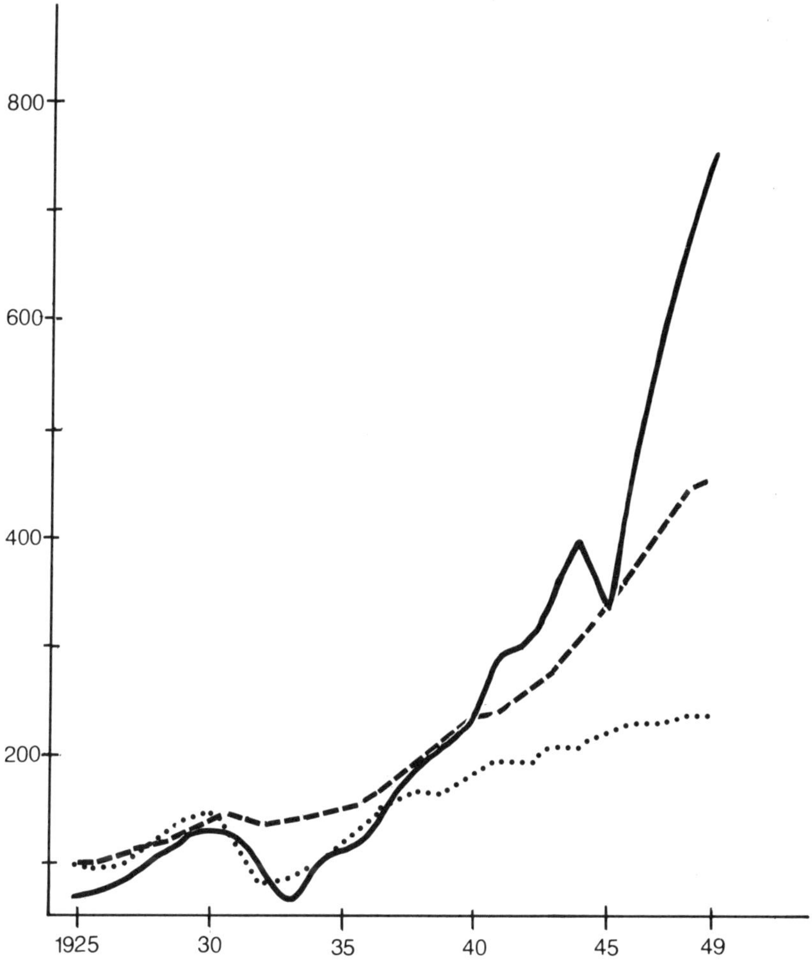

Diagram 3. Shipbuilding.
Production value (—), No. of workers (· · ·) and power installed for direct production (---). Index 1920–1924=100.
Source: SOS Industry.

creased productivity. Nevertheless, investigation shows that this increase in motive power was not so great that it can reasonably provide the whole explanation. The relation between motive power installed and the number of employees appears from Diagram 6. Other factors than motive power must also be discovered to explain the rise in productivity. It is these other factors, the mathematical size and specific economic role of which many authors have tried to assess with varying success, that also include such rationalization measures as the introduction of time and motion studies.

Rationalization measures may be regarded as changes in the composition of contributory factors in production. Above was shown the relation between the human workforce and electrical energy with the idea that the

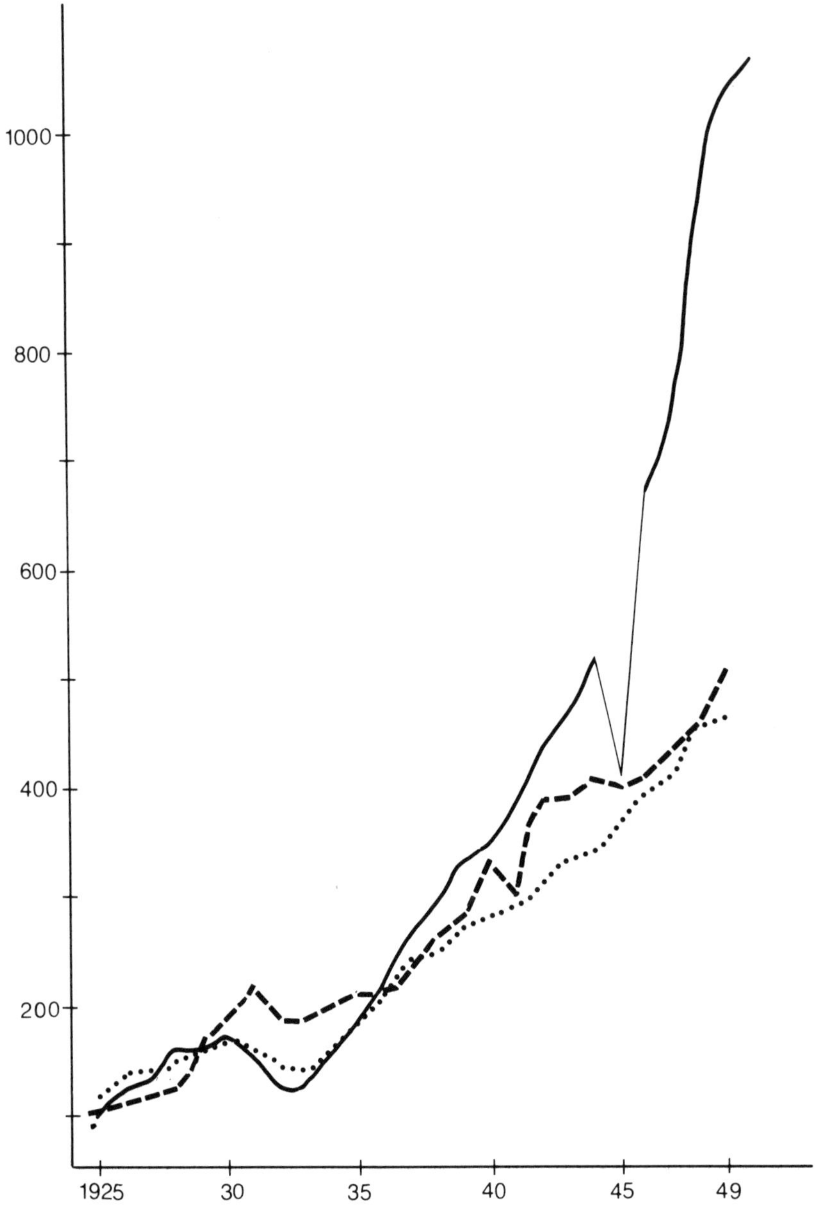

Diagram 4. Electrical engineering.
Production value (—), No. of workers (· · ·) and power installed for direct production (---). Index 1920–1924=100.
Source: SOS Industry.

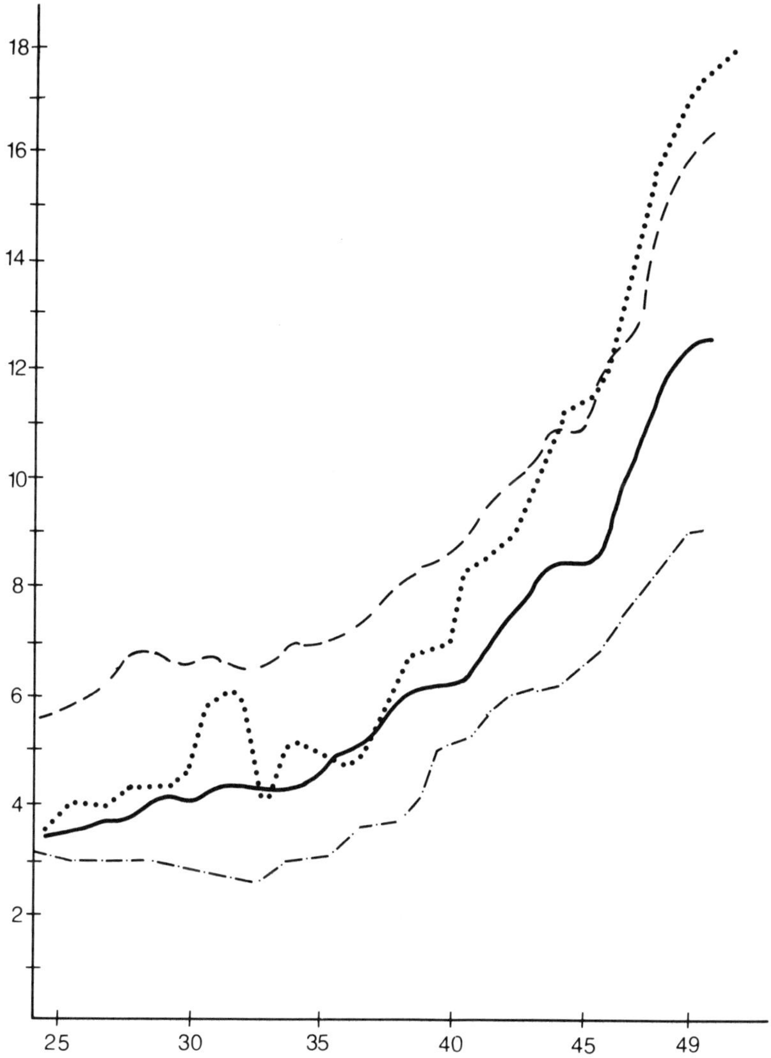

Diagram 5. Productivity in the metal trades industry.
Iron and steel manufacturing (—·—), Metal trades machine shops (—), Electrical engineering (---), shipbuilding (· · ·). Production value per working hour. Skr.
Source: SOS Industry.

development of rationalization carries with it an increased contribution of machine work and a decrease in the human factor in a stated production.

Another change that rationalization is expected to carry with it is a shift in the relation between the worker and the management personnel of the company. The latter category may be presumed to increase in size proportionally to the former. This may be regarded as an expression of the undermining of the skilled worker's role touched on earlier here: engineers take over production planning, economists take over buying and selling,

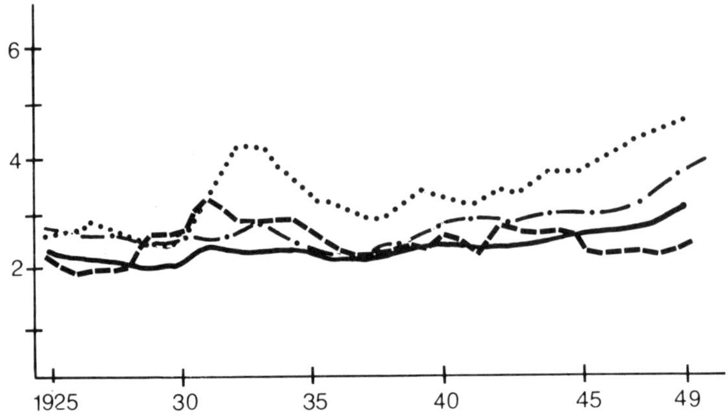

Diagram 6. Mechanization in the metal trades industry.
Electrical engineering (---), Metal trades machine shops (—), Iron and steel manufacturing (–·–), shipbuilding (· · ·). Horsepower (HP) installed for direct production per employee. *Source:* SOS Industry.

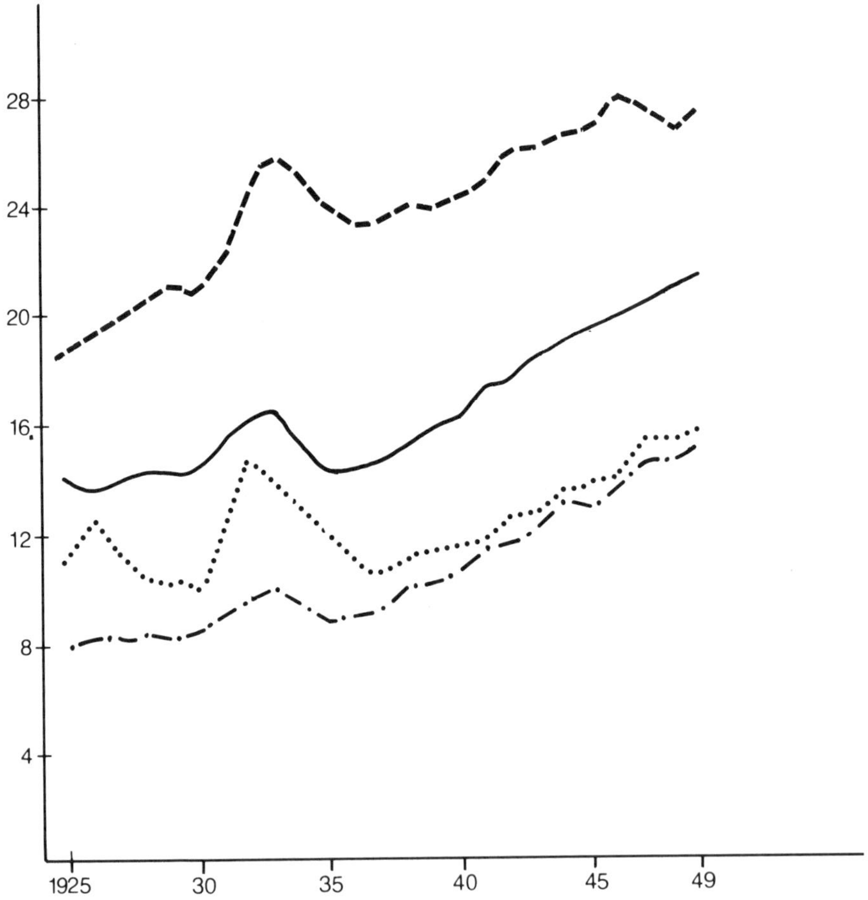

Diagram 7. Metal trades industry.
Proportion of administrative personnel in all employees. Percentage. Iron and steel manufacturing (–·–), shipbuilding (. . .), Metal trades machine shops (—), Electrical engineering (---). *Source:* SOS Industry.

lawyers the necessary work with agreements and contracts. Increased rationalization requires more office staff per industrial worker. Diagram 7 shows the actual position for the selected branches in the metal trades industry.

The differences in levels between the branches is interesting. The modern electrical industry appears as "intelligence-intensive" in comparison with the heavier shipping and manufacturing industries. Otherwise the development is relatively similar in the different trades. The jump in the curves between the years at the beginning of the 1930s reflects no long-term trend but illustrates the effects of the crisis during the 30s. Even though in absolute figures the number of office workers fell because of discharges during the crisis comparatively still more factory workers were discharged, which resulted in the relative increased proportion of management personnel in the total number employed, as is shown. The upward trend seems to be even over the period for the electrical industry whilst for the other three branches' development it seems to have accelerated somewhat in the middle of the 30s. In a comparison with the whole of Swedish industry it can be seen that during the 25 years between 1925 and 1949, the proportion of management personnel increased roughly from 10 to 20 per cent. The electrical industry and also the metal trades machine shops lie well ahead relative to the figures, which was also to be expected in view of the pioneer role these branches had in the industrial rationalization movement.

2. Pay and time and motion studies in the metal trades industry

The metal trades agreement reckoned with two different kinds of pay: hourly pay and piece wages. Since the Metal Workers Union had accepted piece wages in principle, an attempt was made to increase the proportion of work done on piece wages, or the proportion of working hours that were paid at piece wages, the so-called piece wage volume. The size of the piece wage volume in the metal trades industry is given in Table 2. It was, as appears, the jobs in real manufacturing that were put on piece wages; helping and other peripheral jobs carried out by unskilled workers were largely paid at an hourly rate.

The reason the workers' side wanted to increase piece wage volume was obviously that piece wage earnings lay above the purely hourly wages as is shown in Table 3, which is based on the average piece wage earnings. The relative figures also tend to emphasize the pay differences between, especially, the unskilled workers and the other groups.

Pay development in the metal trades plants is shown in Table 4. This states the nominal pay for the different categories of workers included in the agreement. For skilled workers is also given the actual development of pay.

Table 2. *Piece wage volume in the metal trades industry shown as a percentage of total volume of work*

Source: Olsson 1970, p. 74

	1920	1930	1938	1949
Skilled work	76	84	80	81
Semiskilled work	84	85	81	81
Unskilled work	50	56	49	57

The stability of the real wages is striking: a very steady climb with many backward steps periodically during the interwar years was first relieved in the postwar years by a steep rise. The difference in nominal wages between the three categories of skill is also well illustrated, even though the disparity in the nominal values decreases with time.

So to what extent were time and motion studies used in the metal trades industry? An attempt has been made to obtain a picture of this using the Rationalization Commission's representivity investigation. In its approach to 348 industrial companies during 1937 the Commission included a question as to whether time and motion studies were used in the factory and, if so, when these were taken into use.[1]

If mines and purely iron and steel works are excluded there were 81 companies within the metal industries sector that answered the questionnaire. Of these half (40) said that time and motion studies had not been used, 25 confirmed that they had utilized them and 16 did not answer the question, except in one case where the answer was difficult to interpret. The last group may be ignored even though it is obvious that certain companies in it did in fact utilize time and motion studies and actually belonged to the pioneers, as for example Sandvikens Jernverk.

In a comparison between the first two of these groups, those who used and those who did not use time and motion studies, it is clear that the spread across different manufacturing branches is similar. The 40 companies who

Table 3. *The relations between piece wages and hourly wages in the metal trades industry: hourly rates shown as a percentage of piece wages*

Source: Olsson 1970, p. 74

	1920	1930	1938	1949
Skilled work	89	98	95	90
Semiskilled work	93	92	90	89
Unskilled work	89	87	86	85

Table 4. *Pay development in the metal trades industry. Nominal values for skilled, semiskilled and unskilled workers (Skr per hour) together with the real wage development for skilled workers (Index, July 1914=100)*

Source: Olsson 1970, Tables 42 and 52

Year	Skilled worker Nom.	Skilled worker Real	Semiskilled worker	Unskilled worker
1920	2.11	79	1.89	1.72
1921	1.87	83	1.76	1.53
1922	1.28	70	1.19	1.06
1923	1.27	74	1.19	1.07
1924	1.30	76	1.17	1.08
1925	1.33	75	1.17	1.09
1926	1.36	80	1.21	1.11
1927	1.38	82	1.20	1.14
1928	1.40	81	1.23	1.14
1929	1.42	85	1.22	1.16
1930	1.45	89	1.26	1.19
1931	1.44	92	1.30	1.21
1932	1.35	87	1.22	1.13
1933	1.36	91	1.23	1.16
1934	1.37	90	1.25	1.18
1935	1.40	90	1.26	1.19
1936	1.42	90	1.27	1.21
1937	1.44	89	1.27	1.22
1938	1.57	95	1.40	1.34
1939	1.59	95	1.44	1.37
1940	1.69	89	1.53	1.45
1941	1.83	85	1.64	1.56
1942	1.98	85	1.80	1.71
1943	2.04	87	1.84	1.76
1944	2.06	88	1.88	1.80
1945	2.05	88	1.87	1.85
1946	2.25	96	2.01	2.01
1947	2.58	108	2.35	2.24
1948	2.83	112	2.55	2.51
1949	3.01	118	2.71	2.62

did not use time and motion studies were small companies: on the average in 1936 they had somewhat over 142 workers per company. If the two companies which stated that time and motion studies were planned, including Kockums in Kallinge, are excluded from the calculations the picture becomes even more pronounced. Then the companies' average number of workers becomes somewhat over 131. The 25 companies who used time and motion studies had a company average of about 850 workers in 1936. The picture is quite clear: it was the large companies that had the possibilities and felt the need of utilizing time and motion studies.

Table 5. *The use of time and motion studies in the metal trades industry. After questionnaire in the Rationalization Commission's representivity investigation*

1920 or earlier	SKF, Gothenburg
	Separator, Stockholm
	Bolinder Munktell, Eskilstuna
1925–30	Bofors AB
	E. A. Bergs Fabriks AB, Eskilstuna
1930	Norrahammars Bruk, Norrahammar
	Nydqvist & Holm, Trollhättan
1932	Facit, Åtvidaberg
	L. M. Erikssons kabelverk, Älvsjö
1933	Finspongs Metallverk, Finspång
	Atlas Diesel, Stockholm
1934	Skoglund & Olsson, Gävle
	See Fabriks AB, Sandviken
1935	Telefonaktiebolaget L. M. Eriksson, Stockholm
	Åbjörn Andersson AB, Svedala
	AB Sundsvalls Förenade verkstäder
1936	AB Motala Verkstad, Motala
	Sieverts kabelverk, Sundbyberg
No year stated	Eriksbergs Mekaniska Verkstad, Gothenburg
	AB Växlar och Signaler, Örebro
	AB Nordiska Armaturfabriken, Åtvidaberg
	Lesjöfors AB, Lesjöfors
	STAL, Finspång
	Elektriska AB Chr. Bergh & Co, Svalöv
	Köpings Mekaniska Verkstad, Köping

This group of 25 companies will be dealt with in more detail. In Table 5 they are arranged according to the year notified as when the time and motion studies commenced. It is pointed out once again that only a small selection is being dealt with, where the information was easily available, and not any complete list. The main impression is so marked, however, that it is worth stressing. With the exception of certain pioneer companies time and motion studies were taken into use in the metal trades industry at the beginning of the 1930s. They were then mainly utilized by the larger companies.

A more complete picture can be obtained if the Metal Trades Employers Association's internal material is used. In conjunction with the collective negotiations 1931/32 an investigation was carried out by the association's office concerning the position with regard to piece wages. One question asked was whether time and motion studies were used in the calculation of piece wages. 27 companies were shown in the results of the investigation. Of these 16 confirmed that such studies were used. This is a comparatively high figure; on the other hand it is reasonable to presume that the selection was directed at a few, but leading, companies in different districts; the list compiled is arranged geographically. However, comparison of the companies' size produced a picture confirming that given by the Rationalization Commission's material: in the group that used time and motion studies the average company size was 839 workers; in the group that did not the average was only 433.

The most complete picture of the extent of time and motion studies in the metal trades industry can be obtained from a material in an investigation carried out by the Metal Trades Employers Association in 1943. This will be dealt with in more detail.

Together with the association's circular No. 16/1943 was included a questionnaire to members about the piece wage situation. There were 570 answers to the questionnaire from plants employing a total of almost 107 000 workers. Calculated in the proportion of the total number of piece wage hours in accordance with the association's pay statistics the replies comprised 98 per cent. Thus representativity of the investigation was good.

The investigation showed the position at the beginning of the 1940's. But one aspect of the historic development was satisfied by the question of when time and motion studies were first used. In Table 6 the replies are given in a statement of the accumulated values over the period.

In Table 7 are shown further the companies who introduced time and motion studies during the pioneer phase prior to 1940.

Table 6. *The use of time and motion studies in the metal trades industry. Accumulated figures*

Source: Metal Trades Employers Association's special investigation 27/1943

Year	No. of plants	No. of workers	Year	No. of plants	No. of workers
1908	1	716	1928	18	19 596
1909			1929	19	19 655
1910	2	4 669	1930	27	29 341
1911			1931	31	31 414
1912			1932	37	34 093
1913			1933	41	34 523
1914			1934	47	37 630
1915	4	6 165	1935	57	45 350
1916			1936	69	49 463
1917	6	6 664	1937	77	51 731
1918	8	11 175	1938	92	54 883
1919	9	11 311	1939	109	58 363
1920	10	11 903	1940	116	61 293
1921	11	12 029	1941	142	67 880
1922			1942	185	75 284
1923	12	12 162.	1943	201	77 337
1924					
1925			1930/1940	204	81 395
1926			Year not stated	213	82 797
1927	14	14 132			

Table 7. *Year of introduction of time and motion studies (piece wages based on time and motion studies) before 1940*

Source: Metal Trades Employers Association's special investigation 27/1943

1908	Separator
1910	Asea (Västerås)
1915	Sv. Stålpressnings AB (Eskilstuna)
	Sv. Stålpressnings AB (Olofström)
1917	Elektriska Industri AB
	C. O. Öberg
1918	Eskilstuna Jernmanufaktur (Tunafors)
	Svenska Kullagerfabriken (SKF)
1919	Graham Brothers
1920	C. E. Johansson
1921	Eskilstuna Jernmanufaktur (Rosenfors)
1923	Radius
1927	Elektrolux
	Nymans verkstäder
1928	Telefonaktiebolaget L. M. Ericsson
	General Motors
	Thermaenius
	Atlas Diesel
	Original Odner
1929	Archimedes
1930	Volvo/Penta
	Husqvarna
	Norrahammar
	Asea (Stockholm)
	Max Sievert/lödfabriken
	AGA
	Sv. Centrifug AB
	SKF (Katrineholm)
1931	Kabelfabriken L. M. Ericsson
	Asea (Ludvika)
	Stridsberg & Björk
	Monark
1932	Elga
	Volvo
	Svenska Fläktfabriken
	Facit
	Halda
1933	Sv. Skandex AB
	Josef Eriksson
	Strebelverken
1934	Böhlmarks
	Gustaf Eriksson ("Gense")
	STAL
	Elektrolux (Holmsbruk)
	Nydquist & Holm
	Urfabriken, Svängsta
1935	Liljeholmens kabel
	Ulvsunda (second name undecipherable)
1935	Max Sievert
	Bolinder Munktell
	Kockums
	Sv. Järnvägsverkstäderna
	Kolbäcks El. verkstad
	Lidköpings Mek. Verkstad
	Chr. Berg & Co
	Åbjörn Andersson
1936	Tönseth & Co
	Werquist Mek. Verkstad
	Calor
	A. G. Johansson Metallfabrik
	Bultfabriken
	Nord. Armatur
	Skandiaverken (Lysekil)
	Motala Verkstad
	Optimus
	Alnarps Redsk. verkstad
1937	J. G. Ljungman
	Ätranverken
	Rex
	Pump Sep. Gjuteri
	Landsverk
	Wedholms
1938	Helios
	Sv. Radio AB
	Sv. Maskinverken
	Halda taxameters
	Härnösands verkstad
	Köpings Armatur
	Bröd. Forslund
	Hjo Mek. verkst. & El. verkst.
	Norrahammar's other factories
	Åminnebruk
	Smides & Järnsängsfabr.
	Överums bruk
1939	Fundator
	Sv. Bilfabriken
	Sig. Smith AB
	Örebro Bleckemballage
	El AB Skandia
	Arvikaverken
	Jonsereds fabr. AB
	Bröd. Ljungquist
	Nord. Armatur
	Rudelius & Boklund
	Jungner
	Sala Maskinfabr.
	Westerås mask. (Morgongåva)
	Wedevågs Bruk
	Vårgårda Armaturfabr.

During the fourth quarter of 1942, to which the investigation refers, time and motion studies were utilized for the setting of piece wages on an average for 76% of all piece wage time. However, a clear variation is found in various cost-of-living index region groups:

Table 8. *Number of hours on piece wages at plants with time and motion studies shown as a percentage of the total number of hours worked in accordance with the pay investigation.*

Source: Metal Trades Employers Association; special investigation 27/1943

Group	%
I Stockholm	70
II Gothenburg	95
Eskilstuna	80
Malmö	79
Other expensive areas	78
Cheaper areas	74
III	69
The whole kingdom	76

The figure from Gothenburg is remarkably high. Some explanation of this may be that particularly SKF, the Swedish ball-bearing manufacturers, functioned during the interwar years as a training establishment for time and motion study men who later entered Swedish industry. Otherwise the variations are as may be expected: small industry located in the lower cost-of-living areas utilized job analysis to a lesser degree.

From Table 6 it appeared that a total of 213 plants employing 82 797 people used time and motion studies (of the 570 with approximately 107 000 employees who responded) and from Table 8 it is clear that about 3/4 of the piece wage volume was calculated on the basis of such studies. In order to avoid the risk that the few dominating piece wage figures have presented a distorted picture, an investigation was also made to discover the proportion of the *newly-instituted* piece wages that started with the assistance of time and motion studies, Table 9.

What is remarkable here is that at almost half of the plants that actually utilized time and motion studies, these studies were used in 90% of all piece wage setting. In a quarter of the remainder such studies were only used sporadically, in less than a quarter of such setting.

The investigation further showed that 515 job study men were active in industry and that 74 000 employees were affected by their work. At plants with up to about 15 000 employees job studies were carried out by the supervisors. Piece wages were offered directly by the job study men to 27 000 employees, to approximately 10 000 by the manager of the job

Table 9. *Proportion of the newly-instituted piece wages calculated with the assistance of time and motion studies. Accumulated figures*

Source: Metal Trades Employers Association; special investigation 27/1943

Percentages of new piece wages	Accumulated total: Plants	Accumulated total: Employees
100	76	26 177
90	98	38 921
80	108	40 827
70	125	44 013
60	128	44 931
50	144	49 543
25	158	52 789
Lesser figures	185	71 139
Percentage not stated	213	82 797

analysis department, and to approximately 20 000 by the supervisors following proposals from the job analysis department.

The investigation of 1943 was an element in the Metal Trades Employers Association's preparations to obtain a firmer, centralized hold on time and motion studies. Member companies were also asked about the rules and the values used when the results of such studies were translated into offers of piece wages. From the time when these studies began to be introduced during the 1920s there had certainly been some differences in their application, but at the beginning of the 1940s a more stable application appeared. The investigation of 1943 showed that the price per piece (that is to say a price laid down in advance per unit produced expressed in money terms) wholly dominated the setting of piece wages (93 % of all such wages), and that the calculation values such as cash factors, excluded time and lost time allowance comprised common factors.

Gradually norms were established for combining the different factors that had to be considered in the setting of piece wages. A formula was arrived at that was presented to the Labour Court in 1944 and this was included in the agreement four years later. It read as follows:

$A = T \times U \times (1+S) \times (P/60)$, where A was the prospective piece wage item. T represented the time per unit expressed in minutes. There were different methods of calculating it but the most usual was to extract an average value from a large number of job studies. U represented a performance index whilst S represented the so-called lost time allowance about which more is written below. P represented the cash factor, that is to say the standard aimed at for the earnings which would be achieved under normal conditions. As this was usually decided as an hourly earning, the cash factor recurred in the equation which was otherwise formulated in minutes, divided by 60.

The basic idea behind the equation was that a permanent relation existed between time and price: $A=T\times(P/60)$. If the job-operation took 3 minutes on an average and the purported earning level—the cash factor—was 150 öre per hour, the piece rate was thus $3\times(150/60)=7.5$ öre (100 öre=1 Skr).

The setting of a piece wage in this manner, however, would not be correct for a longer working period. A working period also includes breaks, waiting time and a certain reasonable reserve time. Therefore the piece wage formula was adjusted with a special factor, S, a lost time allowance, expressed as a percentage of T. This was a standard allowance which changed the value of time used. If the total time lost, calculated by a standard, is stated as 10% the actual example appears as follows: $A=3\times1.10\times(150/60)=8.25$ öre.

Finally the performance factor was included in order to correct the time and motion study results in the individual case for such effects as depended on the characteristics of the worker who had just been studied, that is to say to normalize or make it possible to generalize the results. A normally good, normally experienced worker received the value of 1. A worker who was more skilled was placed further up the scale: 1.1–1.2–1.3 etc., and a worker who was less skilled received comparatively lower values: 0.9–0.8–0.7, etc.

In order to make this clearer some simple, fictitious examples will be given. Let it be presumed that the cash factor at local negotiations is fixed at 150 öre per hour. For a certain production a time and motion study is carried out on the expert and rapid worker Andersson. After repeated timing it is clear that on an average Andersson requires 3 minutes for the required processing of each piece. Experience shows that lost time for this kind of manufacture may be presumed to be 10%. It is also known that Andersson is clearly more competent than the average worker, thus justifying a high performance index, 1.2. Introduced into the formula the values produce the following piece wage rate: $A=3\times1.2\times1.10\times(150/60)$; $A=9.9$ öre.

The piece wage thus becomes 9.9 öre. If the skilful Andersson proceeds at his working pace for about 50 minutes of every working hour, he can thus make 17 units and thus earn approximately 168 öre at the piece wage. If he succeeds as he becomes more experienced in achieving 20 pieces per hour, he is said to be building up the piece wage. He then earns almost 2 Skr (Swedish crowns), clearly above the cash factor.

The normally expert Bertilsson, who is offered the same rate of 9.9 öre per piece, does not make more than 15 an hour to start with. This gives him an earning around the cash factor 148.5 öre. Following a certain amount of practice he can presumably increase this by some tens of öre. On the other hand the young dreamer Cesarson who regularly seems to have his thoughts elsewhere only makes 12 pieces an hour. This gives him earnings of not quite 119 öre, and not even after long experience does he reach the money factor target.

But what would have been the position if the original job study had been carried out on Cesarson instead of Andersson? Then the working time would have been as much as 5 minutes (60 minutes for 12 pieces). Should no performance index have been used then the piece wage offered would have been calculated thus: $A=5\times1.10\times(150/60)$; $A=13.75$ öre. With this piece wage Cesarson would have, in spite of his slow working-speed, earned the equivalent target. Relatively speaking, Andersson would have been able to make large earnings and after processing collected 2.75 Skr per hour.

In this case, however, it was well-known that Cesarson was not in the habit of working particularly fast; the performance index 0.8 is introduced into the equation. Thus we arrive at: $A=5\times0.8\times1.10\times(150/60)$; $A=11.0$. It can be presumed that a more realistic piece wage has been arrived at, somewhat closer to that which the study on Andersson produced.

In view of this the performance index appears as especially important and around this much of the irritation about time and motion studies was to arise. It was a sensitive matter in many ways. It was an arbitrary evaluation carried out by time and motion study experts, a subjective factor that influenced decisively the "objective" measurement's result. It was in addition an evaluation of a worker's capability, which could often be personally sensitive to the worker concerned. It was quite obvious that this, not least, carried with it strains and stresses on the relationships, both locally and centrally, between the parties of the metal trades industry.

3. The two sides of the labour market

The Swedish Metal Trades Employers Association (Sveriges Verkstadsförening) was established in 1896, primarily on the initiative of the industries based in western Sweden.[2] In 1902 a reorganization produced the form of organization the association was to have during the years with which we are concerned here. In 1917 the association joined the Swedish Employers Confederation. From five geographically separate districts members were appointed to a general council. This organization took care of both the local right of decision as well as providing the possibility of a forceful, central management. The districts collected members' subscriptions and paid over an agreed portion of this to the general council. The latter had to make decisions about such things as industrial action, of which the lockout was considered to be the most important weapon from the beginning. The development towards collective agreements which were nationally applicable and increasing centralization of course shifted the balance towards an increased influence for the general council.

The number of memher companies and of their combined numbers of employees appear from the following diagram.

During the greater part of the period covered by the present investigation *Georg Styrman* belonged to the management of the Metal Trades

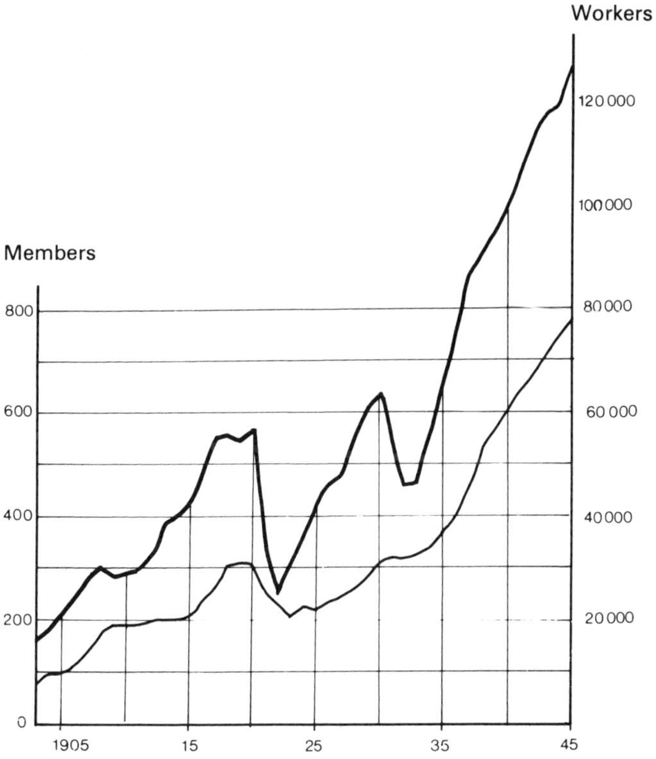

Diagram 8.
No. of members in the Metal Trades Employers Assn. (fine line) and No. of workers employed.
Source: Styrman, 1946 p. 356.

Employers Association. Born in 1883, Bachelor of Law 1910, he succeeded Gunnar Laurell as managing director in 1924. The chairman from 1916–1939 was the leader of Asea, *J. S. Edström.* He was followed during the war years by the managing director of Husqvarna, *Gustaf Tham.* From 1946–1949 the chairman was *Arvid H. Nilsson* of Eskilstuna.

Between the monthly meetings of the general council an executive board dealt with current matters.[3] There were also various committees functioning at the central level. Of especial interest for this investigation is what was called *The Apprentices Committee,* constituted to take care of a limited task, but which was enlarged from time to time to embrace more general technical matters, not least the various problems associated with time and motion study. For this purpose there could also be appointed a so-called technical committee when required. Worthy of special mention here is *Erik August Forsberg's* role as leader of *the Committee for time and motion studies,* emerging in 1941 from the Apprentices Committee. In addition to this basic structure of the organization there were as a rule special negotiat-

Table 10. *The Swedish Metal Workers Union: number of members and number of branches*

Source: The History of the Swedish Metal Workers Union, parts II–IV. Table 1

Year	Members	Branches	Year	Members	Branches
1920	69 786	251	1936	129 331	273
1921	62 357	249	1937	146 816	294
1922	54 837	244	1938	153 531	319
1923	55 331	245	1939	165 049	329
1924	63 361	244	1940	175 131	337
1925	70 998	239	1941	181 322	338
1926	76 591	243	1942	190 777	338
1927	80 999	244	1943	199 216	343
1928	89 019	242	1944	202 090	346
1929	98 131	242	1945	198 715	350
1930	101 580	246	1946	202 420	348
1931	101 439	253	1947	212 000	350
1932	100 666	253	1948	225 035	353
1933	101 021	253	1949	215 535	352
1934	109 068	260	1950	220 114	353
1935	119 903	266			

ing teams appointed for the negotiation of new collective agreements at the national level.

The Swedish Metal Workers' Union (Svenska Metallindustriarbetarförbundet) was established in 1888 and was affiliated to the Swedish Trade Union Confederation (Landsorganisationen, LO) in 1904.[4] The union's organizational growth during the period concerned appears in Table 10.

The first national agreement for the Metal Workers Union came in 1905. It included provisions about piece wages: from the beginning of the mid 1890s the union had started to relax the completely negative attitude towards the piece wage system that it had had earlier.

The supreme organ for decisions within the union was Congress, which assembled every fifth year with the participation of representatives for all branches. From the end of the 1920s policy conferences to discuss forthcoming collective negotiations were called regularly with some hundred representatives from the constituencies into which the union's branches were divided. These policy conferences also had an advisory function on questions concerning whether the agreement should be terminated.

At the congresses a general council was elected, which had up to some 30 members for the appropriate period. Of these about 7 were elected to the executive board.[5] Like their counterparts the union also used to appoint special negotiating teams for the negotiations on collection agreements.

The list of union leaders in the Metal Workers Union is most impressive. Many of these were also to be recruited to still more responsible functions in the workers' movement.

Johan-Olov Johansson became a duly elected representative (ombudsman) in the union as early as 1908 and its chairman in 1920. Five years later he was elected to the post of treasurer of the Swedish Trade Union Confederation. He was followed by *Fritjof Ekman,* ombudsman with a radical accent since the beginning of the 20s. Ekman led the union until 1932 when he was sworn in as Minister for Commerce and Industry in the Social Democratic government of Per Albin Hansson.

Following Ekman came *Gunnar Andersson,* who only four years later transferred to the post of vice-chairman of the Trade Union Confederation. Ten years later he was elected to succeed August Lindberg as chairman of the Trade Union Confederation, but died shortly afterwards. In 1936 Andersson was replaced as chairman of the Metal Workers Union by *Oscar Westerlund,* who had been vice-chairman since 1922 and who had competed in vain at the elections to the chair in 1925 and 1932. Presumably his political affiliation had been an obstacle: he was not a social democrat but belonged to the socialist party, a communist party not belonging to the comintern. There was some stir when he left this party for the social democrats in 1931; Westerlund had been the foremost trade union functionary in the socialist party. He was the union's chairman for twelve years and thus had the task of fighting the hardest battles with the communists within the organization in the metal workers' conflict of 1945. In 1948 he was succeeded by *Arne Geijer.*

Along with the Metal Workers Union there was also a special union, the *Swedish Foundry Workers Union* (Svenska Gjutareförbundet), which conducted its own negotiations with the Metal Trades Employers Assn., usually parallel with the Metal Workers Union. The Foundry Workers Union will be mentioned sometimes in the following pages but does not play any greater part in the present investigation. Nor do the unions that organized the foremen and office workers in industry.

4. The Swedish Metal Workers Union and rationalization

The Metal Workers Union's attitude to rationalization in general—which of course also constituted a condition of the organization's attitude towards job studies—was expressed in a series of statements. The course of these was marked by rivalry between the reformistic union leadership and the more radical and especially communist opposition. The situation was similar in the Trade Union Confederation (LO). There also statements on policy were forced out in argument with those more radical. As the Trade Union Confederation's attitude can be said to lay down the lines for the trade unions it is of interest here to show some of the steps in the Trade Union Confederation's development.[6]

At the Trade Union Confederation's congresses of 1922 and 1926, the executive board laid down that it was the general economic conditions, improved methods of production and more effective economic organization that provided the conditions for successful union activities. This indicated a basic assessment of rationalization without specially emphasizing the trades union organizations' right to insight and joint determination. At the meeting of the International Confederation of Free Trade Unions in Paris in 1927, the positive speeches about rationalization made at the World Economic Conference in the same year were approved. The importance of the workers obtaining influence over changes in production and sharing in the profits was however emphasized more strongly. The positive line, which became that of the Trade Union Confederation during the Industrial Peace Conference of 1928 (a Swedish equivalent of the Mond-Turner talks in Britain) appears as a natural and consequent result of the unions' behaviour in the international arena. Even in the subsequent industrial peace delegation's report of 1930 the questions of rationalization, influence and sharing are linked in a similar manner.[7]

The Trade Union Confederation's successive attitudes were accompanied by a radical, obviously communist, opposition. At the confederation's 1922 congress, the demand was made for a stronger workers' organization, in opposition to the leadership's emphasis on general economic conditions being decisive for the result of collective negotiations. At the 1926 Congress the question of centralization was again brought up, this time through the action of the communist-dominated Branch 41 in the Metal Workers Union under the chairmanship of Ernst Ohlsson. The same branch had earlier during the year been behind what was called the Gothenburg conference.

In 1925 the leaders of the Trade Union Confederation had ignored a request from local branch No. 41 of the Metal Workers Union to arrange a uniting trades union conference. In opposition to the leaders' indifference the branch supported the conference, which was held in January 1926. This was an expression of the communist attempts to create a homogeneous trade union movement where it would be possible to assume the leadership. This was to occur by making radical and concrete daily demands that would appeal to the social democratic workers/voters and thus compel the social democratic political leadership to adopt the communist line. The trade union demands were thought of as means to achieve a communist-led political workers' movement.[8]

The Gothenburg conference had not adopted too defiant a tone towards the Trade Union Confederation and the social democrats because the united front's softer tactics were to be followed.[9] But it was against a purposeful opposition—the communists behind the Gothenburg conference—that the line in favour of rationalization was formulated at the confederation's congress in 1926.

In 1928 the radicalization of the Red International of Trade Unions (Profintern) took place and the united front committee set up by the Gothenburg conference followed the new signals. The material for a more powerful offensive lay close at hand: the miners' strike, the legislation on collective agreements and the Labour Court, and also the social democratic unions' collaboration in the industrial peace conference. The united front committee issued invitations to a new national Trade Union Conference, a rival conference to the government's, in January 1929. The national conference producted an action program for the trade unions, the first point dealing with the struggle against capitalist rationalization.[10] The powerful opposition of the united front committee brought a counteroffensive from the Trade Union Confederation's leadership during the spring of the same year and compelled the united front committee to retreat.[11]

At the Trade Union Confederation's Congress in 1931 the organization assumed a still clearer attitude on rationalization. Congress approved a statement by the International Confederation of Free Trade Unions where the positive basic idea was emphasized at the same time as guarantees were required against such misuse of rationalization as could be to the detriment of the workers. The demands for influence and participation in the profits of rationalization were put forward once again.

Against approval of the International Confederation of Free Trades Unions' statement there was a more radical proposal by a representative of the Typographers Union. The crisis of capitalist society was identified and judgment on rationalization was summed up in the following passage:

> Every means that capitalism has attempted to use for solution of the crisis has been transformed into a scource for the working class. Capitalism has shown itself incapable of organizing trade and industry and administering the values they have seized in such a way that they serve the public welfare.[12]

Congress decided, without any real debate, to reject the proposal.[13]

But it was not only by the International Confederation that the problem of rationalization was raised. Even on those points in the agenda of the meeting dealing with working hours and unemployment, rationalization was used as an argument, mainly in motions from the communists.

Rationalization was also taken up as an independent point. Branch 41 of the Metal Workers Union in Gothenburg proposed that an investigation should be carried out as to whether to the national secretariat should be linked a special organ "for the carrying on of studies concerning development of the work process ... and for the initiation of information activities from the movement in these questions".

The statement, which was adopted by Congress, with communist opposition, received but little debate, and once more confirmed what was now the customary position: rationalization was inevitable and unavoidable, but could be accepted only under certain conditions:

> Even though the position is such that the character of rationalization is economic, that is to say the greatest possible production results shall be achieved at the lowest possible costs and with the least possible investment of capital, if it is to be regarded as at all rational in any real meaning of the word, the purpose of it may not be entirely economic but rather social. In other words rationalization must constitute a

path in the struggle to promote the common weal and to raise the economic and social standards of the whole people. The reduction of production costs by means of technical and organizational measures in order to improve and develop production and distribution of necessities may in itself be praiseworthy but only become of any real meaning to the degree that occurs with consideration to the general prosperity of society.

Here the societal aspects are emphasized rather differently from the International Confederation's statement that had been accepted by Congress the previous day. However, this perspective shifted the problem from the immediate horizon of the Trade Union Confederation. The local organizations were given a controlling and preventive function but no tasks for the Trade Union Confederation or its secretariat were mentioned. But what it was possible to do in addition in order to prevent mistakes in rationalization or other indications of misdirected private capitalism lay within the field of the political organizations. So the problems were moved from the Union Confederation to the unions, associations and branches and to the political section of the workers' movement, especially the Social Democratic Party.

This policy was brought forth under a continuos fencing with the radical, often communist opposition. In the communists' tactics to create a profile for themselves in the current problems with radical demands against the social democrats, questions about rationalization formed one of the main sources of argument. This in its turn compelled the social democratic majority and the leadership to assume positions that usually defended rationalization even within the framework of capitalist society, under the assumption that certain conditions were fulfilled. These statements created the conditions for an approach between the trade union movement and the employers: a pronounced common view of the importance of measures for the furthering of production created a basis for a community of views and exchange of ideas. It was an advantage in the cause of industrial peace and the growing cooperation between the employers' organizations and the unions but presumably also created problems for the social democratic leadership in the Trade Union Confederation. There is no doubt that rationalization arguments were potentially powerful in the communist propaganda. There were many unemployed during the interwar years, reductions in piece wages and demands for increased tempo were also realities. A trade union program where it was possible to integrate rationalization, with its advantages but without many of its negative results, should paralyse the force of the communist reasoning.

It was necessary for the Metal Workers Union to make its decisions in harmony with the development described in the Trade Union Confederation. There also it was a case of a struggle between a reformist organization's leadership and an opposition made up of radical groups. The communist metal workers' role in the course of events in the Confederation has already been described. Quite obviously they behaved in the same manner in their own union. Needless to say, the great Metal Workers Union was one of the most important strongholds to conquer for the trade union opposition of the communists. The struggle between a more radical swing to the left and the restraining aim of the majority runs like a main thread through all of the union's business. At the beginning of the twenties the struggle was about factual issues such as affiliation as a body to a political party, the right of decision on agreements and control over the conflict weapons. Around 1925 the settlement of the winter's conflict caused irritation between the union's leadership and the leaders of the communist-dominated Gothenburg branch. The Gothenburg conference in the spring of 1926 also led of course to great antagonisms. In most questions these penetrated into the union's inner circle: the communist Oscar Westerlund had been vice-chairman of the union since 1920 and the personal attacks in conjunction with the election of chairman in 1925 and 1932 were extremely bitter.[14]

The formations were thus well trained at the Metal Workers Congress in 1929 when the union was to decide about the more open struggle, which the so-called national conference had emphasized in the spring. (Compare p. 49) A conference program was presented to Congress in a motion from Branch 18 in Arvika. The pith of the motion required that the union should join the united front.

The general council dealt with the Arvika motion together with a much shorter one on the same subject from Kalmar and also a motion from the Gothenburg branch dealing with the shortening of working hours. It says a great deal that they were included in the agenda of the meeting under the heading of "Tactical questions". It was more a case of tactics than of facts. The council was compelled by the arguments on the motions to make a decision in principle about rationalization:

> With reference to these motions the general council would like to state that the union within its field of activities ... has always vigorously maintained, that in Swedish industrial life the best and most modern methods of production and auxiliary devices must be used. The workers demand that modern machines and the best possible tools shall be used in order to facilitate and improve the work process so that better and cheaper products are produced than with out-of-date technical equipment, at the same time the workers, because of this, should be able to enjoy higher pay and shorter working hours.
>
> ...
>
> The exhortation, however, which in motion 3 (the Arvika motion) calls for a struggle against 'capitalist rationalization' appears to the general council to be a call to depart from the main direction the union has followed up to now in its political wage-activities. The council is unable to recommend the acceptance of such a request to Congress, as it is hardly possible to differentiate between a rationalization in its usual meaning and what the movers of the motion call capitalist rationalization. Should Congress decide in accordance with the requirements of the motion and the decision should be possible to carry out in practice, it seems to the council that this would be a case of attacking the future development of the production process, which is necessary in order to produce still better and cheaper products, and thus make more difficult a continued improvement of the members' standards of living. Instead of fighting rationalization in principle, the union should be concerned with gaining what is possible for the members in the wage policy area that which is

possible of the better results of production that rationalization should bring about. With regard to the detrimental effects of rationalization, such as an increase in the number of sickness and accident cases, it is the union's job in consultation with the workers' representatives in the Riksdag to press for fully comprehensive legislation in such matters.[15]

In the debate which followed the union's chairman, Fritjof Ekman, gave further interpretations of his view of rationalization and the union's role:

What is always present for the union in wage negotiations is the limitation of capital's share in the results of production to the benefit of the worker, and to act so that the workers also may obtain influence in the process of production.[16]

The principal attitude for which Fritjof Ekman pleaded is recognized. To this he added a "sensible and practical" attitude to questions about pay and working hours, which left plenty of room for attention to be given both to the economic position and the employers' actions when formulating the workers' current demands.[17]

The most powerful arguments for the opposition's point of view were put forward by a deputy from an iron works in Uppland:

When Ekman says he has difficulty in understanding capitalist rationalization, I understand him well. Ekman has not worked in any plant in the last few years and therefore cannot understand how conditions are nowadays. Up in Uppland iron works with 200-year traditions have been closed because of the results of rationalization. A horrible misery exists among the unemployed workers from these ironworks. If Ekman has, as I have, seen these results, he would have had the same idea as I have about capitalist rationalization. If he comes up to the workers at the ironworks in Uppland, these descendents of the old Walloons, and he develops his rationalization program there, he will surely find it difficult to get them to understand it. . . .

There is a great deal of difference between our old employers and the management we now have at the ironworks. Now it is usually a bank that directs the company and in its service there are special office workers who only try with one method more rational than the other to get the favour of their taskmasters. If the workers land in misery this does not trouble these gentlemen. . . .[18]

Even Oscar Westerlund defended the line of argument of the motion, but in somewhat softer phrases. He produced objections to rationalization derived from experiences in the USA. The workers had not obtained a higher standard of living, Westerlund stated, the high pay policy applied mainly to the building industry and rationalization had forced up the tempo of work. But the conclusion was:

For us to stand out against rationalization is not possible. It might perhaps mean putting obstacles in the way of development, but we could ensure that we counteract its worst effects.[19]

Congress rejected the motion's demand and accepted in all respects the majority line of the general council. But Congress adopted one demand presented during the debate, namely that the council be instructed to present to next Congress an action program for the union on a number of current questions.[20]

As an element in the work involved can be seen the general council's discussion about rationalization in November, 1930.[21] The debate was preceeded by information on the subject and Tarras Sällfors of the Industrial Office Ltd. (Industribyrån AB, a consulting firm owned by the Federation of Swedish industries) had been invited by the council. His talk on "Rationalization and its importance" was met with approbation.

The discussion within the council, dealt with in detail by Back, was mainly a demonstration of well-known views. It was based on foreign experiences. However, from Sweden there was no material about rationalization.[22]

Fritjof Ekman was undoubtedly the most positive to rationalization and appeared almost enthusiastic about the possibilities of time and motion studies:

> The trade union organizations cannot object to the purpose of rationalization. For the purpose is to arrange production in the most sensible and practical way in order to achieve higher standards of living and shorter working hours. It is possible to argue about how the workers should react in detail at the various work places. By looking ahead, this question winds up the whole pay problem. The situation is that by means of time and motion studies the work is confirmed at its real value, but this presumes that all of the factors that achieve this measurement of value continue to be present. The setting of piece wages by means of negotiations cannot achieve a correct evaluation. . . . I am not in agreement with those who advance the view that rationalization causes unemployment. On the contrary, unemployment would be even greater if rationalization is not undertaken. . . . Naturally, within the various companies the workers should point out the measures that must be undertaken if there is to be any meaning at all in time and motion studies . . .

A glimpse is caught here of the workers' positive, active collaboration in rationalization.

Proposals were presented to the council to set on foot in various ways further investigation of the rationalization question and by means of articles and circulars to give the members instructions on the attitude they should take with regard to rationalization. Fritjof Ekman, however, referred to the statement on this matter that had been approved by the previous Congress and considered that the discussion carried on could constitute sufficient guidance for the union's continued action. The general council agreed the chairman's line with a certain amount of dissension (14 against 11).

This decision meant that the union's line had withstood still another test after having been confronted with a detailed and authoritative presentation of rationalization's methods and theories. Certainly one of the speakers had pointed out in this connection that there of course existed a great deal of difference between theory and industrial practice. Nevertheless the line adopted by Ekman and the majority of the general council had once again been confirmed.

This was also one of the grounds for the debate on rationalization at the union's 1932 Congress where the battle lines were again to be drawn up between the social democrats and the communists. The battle had flamed up once again in conjunction with the engagement in the industrial peace delegation and the industrial peace committee following the Industrial Peace Conference of 1928 (cf. pp. 48, 258) and increasingly so following the incidents in Ådalen (one of the more dramatic events in contemporary Swedish history: five persons were killed when military forces opened fire against workers in a demonstration against strikebreakers). Many of the motions placed before the Congress in 1932 were of an ideological character—this was a general attack.[23]

Rationalization entered into the debate on many factual issues. A list of motions demanded that rationalization and unemployment should be used to justify a demand for shorter working hours. Congress followed the general council's recommendations, which made a weak admission of a possible connection but confirmed that assaults on the working hours question should mainly concern members who had not yet obtained the 48 hour week.[24]

The rules for conflict pay were also examined by Congress. Branch 41 had requested that the union should support members who had gone from piece wages to hourly pay in order to blunt one of the employers' systematic attempts to lower the wage level by lowering piece wages. The majority of the general council had approved of the demand made in the motion but Fritjof Ekman made a reservation. He presented his argument in a forceful speech at Congress in which he combined practical and theoretical legal arguments. Following a long and tough debate, where the views on the part rationalization should play once again were great importance in the time-honoured way, Ekman's amendment was carried by 102 votes to 86.[25]

So again, through these decisions, the line favouring rationalization was confirmed in spite of negative effects such as unemployment and increased working tempo. The means of rationalization were also dealt with by Congress. A motion compelled the general council to make a principle decision. The motion said that of latter years employers had started to use scientifically constructed time and motion studies and went on to say:

> From the workers' point of view we have nothing against time and motion studies if these are carried out with complete objectivity and are decisive. As the position now is we are completely without any knowledge of both the calculations and the grounds for calculation and are quite unable in any way to check the results of time and motion studies, which for us is an absolute necessity if we are to obtain any part at all in the profits of rationalization. It seems impossible for both the local branches as well as their committees to safeguard the members' interests in a satisfactory manner and we therefore propose that Congress agrees that:
>
> The general council must pay the greatest attention to rationalization and associated questions such as time and motion studies, changes in working methods, etc., and in order that this may be done in a satisfactory way the union should appoint a technically trained person who is qualified to hold courses of instruction and by means of lectures and technical journals supply the members with the essential knowledge necessary for us to have in these matters, and also to assist at negotiations where a technical expert is desirable from our point of view.[26]

In its reply the council confirmed that it was in touch with industrial development and thus also the problems of rationalization. The introduction of time and motion studies carried with it certain consequences. The council continued:

> The procedure in the setting of piece wages has been changed a great deal by these in some industries. The method used earlier, where the worker and foreman or the supervisor involved agreed about the piece wage, has been relinquished in the wellorganized and specialized industries as it has not been considered satisfactory or suitable. Planning and costing departments has been set up instead with the task of planning and carrying out the work in the best way and in conjunction with this, including the use of time and motion studies, to calculate the piece wages. The purpose of this should be by means of careful and exact investigations, to discover how much can be carried out in a certain period with the tools available. The investigations thus carried out then form the basis for the assessment and setting of the piece wages. This is an objective method of procedure in the respects mentioned, and as the results of it may be decisive in the setting of piece wages the general council considers, as the proposer of the motion, that it is unable to raise any objections.[27]

The position taken here, not only with regard to rationalization but also on time and motion studies, thus left no room for doubt. In line with this the general council assessed the workers' rights in the setting of piece wages and their chances of asserting those rights. There were provisions in the agreement that the piece wage should be arrived at by negotiations, said the council, and if these were claimed consistently many sources of dissatisfaction should be eliminated. That a technically trained person attached to the leadership should be of any use in this context was rejected with arguments that also set a limit to the union's ambitions to act:

> The grounds for the setting of piece wages must so far as possible be laid down in agreements and with the guidance of the agreements' contents and local conditions it is up to the members in every locality to assess the jobs and try to come to an agreement on the piece wages. For an outsider cannot be in a position to assess the technical possibilities of the performance of the work in question and as a result of this he is also unable to decide whether the piece wages are correctly assessed or not.[28]

Without further debate Congress adopted the council's proposal to the effect that the general responsibility of the council to continue to watch developments stood fast, whilst the proposal for definite measures—to engage an expert for information, instruction and to assist at negotiations–was rejected. Because of the views thus put forward by the council about the central organization's chances of dealing with piece wages, it followed that efforts were directed to questions concerning the collective agreements, which was where the union's efforts could be made. At the same time it is clear that the union's general council was not prepared to accept the additional responsibility of rationalization questions. As with the leadership of the Trade Union Confederation, the general council of the Metal

Workers Union passed the responsibility further downwards in the organization.

5. Development of the metal trades industry's agreement

The concluding section in this presentation of the metal trades industry's conditions will be devoted to a survey of changes in the paragraph on piece wages in the metal trades collective agreement, and this will be the central point of interest in the continued investigation of the collective negotiations. Certain extracts from the agreements' texts are given in extense in Appendix 1.

The paragraph dealing with piece wages, in the metal trades collective agreement, which in the agreements of 1919, 1923, 1924 and 1925 bore the serial number 11 and in the following agreements number 4, mainly had the following arrangement and contents during the period investigated.[29]

Clause 1 regulated the application of piece wages and laid down that piece wages should be used whenever possible (prior to 1929 "may be used").

Clause 2 regulated the occurrence of fixed piece wage schedules with definite periods of validity. Its significance was that in repetitive manufacture the piece wage should be bound so that it could not be arbitrarily altered except when important changes in the working methods or machines were made. The fixed schedules played a lesser part in the metal trades industry than the free piece wages.[30] The men in the plants were unwilling to pin themselves down and an increasing tempo of rationalization provided such great possibilities for the employers to refer to altered conditions of production that this section became of declining importance.

Clause 3 dealt with the free piece wages and laid down that agreements on these rates should be made at free negotiations between the employer or his representative on the one hand and the worker or workers who were offered the wage on the other. Normally agreement should be reached before the work commenced but afterwards in certain cases if the parties were so agreed. In the agreement of 1936 an addition was introduced to the effect that a piece wage that often recurred, which had been set through time and motion studies or had been controlled in some other way, should, if the parties were in agreement, be included in the plant's fixed piece wage system and have the same protection as the fixed schedules in accordance with clause 2.

Clause 4 applied originally to the bonus piece wage rates and laid down that the grounds for these should be made the subject of negotiation and agreement. The clause was similarly worded in the agreements from 1919 to 1935, but a series of changes was initiated in 1936 that made the clause the focal point of this investigation. (The wording of the clause is given in Appendix 1).

The collective agreement of 1936 thus provided that workers offered a piece wage based on time and motion studies should have the right to know the grounds on which the studies were carried out and the results to which they led. The agreement that followed in 1938 used a looser terminology "the workers", which could be so interpreted that the right applied not only to the workers who were to have the piece wage but also the organization, in those cases where there were disputes about the piece wage. This meant that the door was slightly ajar for the organizations' insight and influence in the setting of piece wages. Furthermore the 1938 agreement contained the alteration that the workers not only had the right to know the basis for time and motion studies and the results but also to examine the report on time and motion studies in every individual case.

During the early years of World War II no change was made in that clause of the piece wage paragraph but changes were made in the 1944 agreement and strengthened the workers' position. It was now stated that even the representative of the local trade union was entitled to the same right as the workers directly affected to examine the time and motion study system and the report. This implied a stronger position for the local trade union organization and was also strengthened by the right to request and follow a control study in the event of a dispute about the studies. Additional minutes confirmed that the workers had the right to take the initiative for time and motion studies and to have a contributory influence on the extent of the studies in the particular case concerned. To this was added a recommendation to negotiate and make agreements on the grounds for calculation for piece wage setting with time and motion studies. On these points the employers had been compelled to admit the rights of the workers. As a counterbalance may be regarded the fourth minute, which accentuates the demand for loyal cooperation in correct studies.

Following the great strike of the metal workers came the agreement of 1945, which brought with it changes in the minutes on clause 4. The most important of these was the setting up of a common central job study council. This created a tool with which the organizations would be able to deal with the complicated issues of principle concerning job studies. The institution of the council made it still clearer that problems about time and motion studies were not something exclusively for the employer but for the organizations of both parties. At the same time this council emphasizes a clearly common fundamental view. The council aimed at joint, unanimous decisions in accordance with intersubjective norms (objective norms in time and motion study methods), not horse-trading between the employers and employees. However, the possibility of the latter was retained in the last sentence of the minutes.

The agreement of 1947 brought with it a sharper tone in the minutes for the purpose of compelling local agreements on the bases of calculation (minute 3) and also to prevent the employers from refusing, with reference to competitive secrecy, to hand over the job study reports to representatives of the Metal Workers Union.

As has appeared, paragraph 4, clause 4 of the metal trades agreement during the decade from the middle of the 1930s to the middle of the 1940s, was enlarged in such a manner that its main substance was subjected to a change from a regulation of the increasingly more rare bonus piece wage to a regulation of the application of time and motion study methods. Every change marked an increased or more assured influence for the workers.

Clauses 5–8 in the agreements up to and including 1947 did not particularly affect, and were not particularly influenced by, the introduction of time and motion studies. Neither, with certain exceptions, were changes made to any great degree during the period. In the 1948 agreement these clauses appeared as numbers 6 to 9.

Something new was introduced in the agreement of 1948 as clause 5. The text of the agreement is given in Appendix 1. This clause contained a standardization of the method of calculating piece wages from time and motion studies; the formula for calculation of piece wages, presented earlier, was reproduced and the cash factor expressed in öre was given for each cost of living index region.

Local freedom of action was thus curtailed for both parties, this marking the end of the employers' view of time and motion studies and the calculation of piece wages as a field beyond the reach of the trade union organizations' claims for a contributory influence. At the same time the clause showed that the workers' trade union organizations accepted time and motion studies and on them based the setting of piece wages with all the associated technicalities.

The above survey has shown that the kinds of rationalization of which time and motion studies were an important part influenced and were reflected in the piece wage paragraph in clauses 4 and 5 of the 1948 collective agreement of the metal trades. The most important changes are to be found in the agreements of 1936, 1938, 1944, 1945 and 1948.

Every change noted meant a shift in the balance of influence between the employers and the workers' trade union organization. The setting of piece wages, which during the mid-1930s solely concerned the employers and workers directly affected, became gradually and increasingly the obvious subject of agreements with the trade union representatives or were subjected to their approval. From the beginning, the employers had stressed that time and motion studies were a domestic matter for every company. During the period covered these studies became increasingly a matter for the organizations on both sides of the fence. Thus the introduction of time and motion study methods may be said to have influenced the balance of the labour market in two ways. The employers' dominance and influence was slightly reduced and the organizations of both parties increased in strength in relation to their members.

CHAPTER 3

The Metal Workers Union and the pay guarantee: the Chairman's line

1. Introduction. The collective negotiations of 1928

In the previous chapter a survey was made of the actual changes in the piece wage paragraph of the agreement of the metal trades industry, which can be said to have been caused by rationalization methods and especially by the introduction and designing of time and motion studies. Here and in the following chapters attention will be given to the intentions, desires and motives that led to these changes.

The changes due to the experience obtained from time and motion studies started at the beginning of the 1930s. Every alteration in the text of an agreement is the result of a process that may have been instituted very much earlier. A certain degree of conservatism is inherent in every collective agreement. One knows what one already has and every demand for a new factor to be introduced usually has a long time in which to mature. The demand has to be given a formal shape that will answer its purpose and be resistant to exploitation by the other party. It must be delicately weighed and given priority in relation to other wishes and demands that may exist and it is almost an unwritten law that new demands frequently have to make way for those which are older. In the changes in the piece wage paragraph of the metal trades agreement the maturing process in so far as it can be traced in the central material took about ten years, which takes us back to the middle of the 1920s.

From 1 January 1925 there was a one-year metal trades agreement.[1] In the autumn of the same year the Metal Workers Union's executive board convened a policy conference to discuss the agreement for the workers of the metal trades industry, which was to decide whether notice should be given to terminate it.[2] With a 2/3 majority the conference accepted the suggestion put forward by the executive board that notice should not be given in view of the fact that the trade situation was unfavourable. The agreement was extended.[3] History repeated itself the following year when the question of notice was taken up at the union's regular Congress. On this occasion differences of opinion among the representatives were more marked and in the discussions there were signs of dissatisfaction with the principles used for the setting of piece wages.[4]

In the autumn of 1927 assessments of the economic situation were somewhat rosier in the Metal Workers Union's management, but nevertheless the executive board led by Fritjof Ekman recommended that notice of termination should not be given but that the energies of the union should be concentrated on fighting the liberal government's plans on the question of industrial peace. An eloquent opposition mainly stemming from Gothenburg and Västerås made itself heard and the demand for termination was advanced in order to push the questions of minimum pay and holidays. However, the executive board was unanimous in its recommendation even though there was one difference of opinion that arose on the question of motivation. The vice-chairman, Oscar Westerlund, thought the economic outlook was not quite so bright as the chairman had pretended and the future did not look promising, not least because of the threat of rationalization to the workers.[5] The debate turned into an alteration about various views on rationalization. But Ekman did not share Westerlund's assessment at all:

> Rationalization should not be opposed as it is a condition of development and for a greater production of necessities. It was no tragedy for mankind that, for example, spinning machines, steam engines and the motor were invented and adapted to the service of mankind. These inventions, as with others, have stimulated development and contributed to the increase of public welfare.[6]

The divergence of opinions in the Metal Workers Union between two different ideas about the value of rationalization to the workers has been touched on before here. Ekman's positive stand may to a certain extent explain his view of the demands that were put forward about changes in the provisions on piece wages. He admitted that a definite margin of income for piece wage work (for example calculated as an extra percentage on the hourly wage) was an old demand from the union but he considered it a waste of time to start a battle with the employers about it. Such provisions were not to be found in any agreement and were in opposition to a principle view of the employers. Satisfaction with the setting of piece wages depended to a certain extent on the fact that the workers themselves had been unsuccessful in making use of the protection already provided by the terms of the existing agreement. Ekman continued:

> The calculation office, which exists in a number of factories, is certainly not supported by the agreement, but neither is it possible to deny the employer the right to submit recommendations about price. . . .[7]

With this the new reality the growth of time and motion studies and factory organizations set up for these had made its first impact in the workers' discussions about the agreement.[8] The reaction derived from a positive view of rationalization and time and motion studies were understood to be one of the employers' justifiable tools when the piece wage was to be proposed.

The conference followed the executive board's line with regard to notice of termination; but a proposal for a new agreement was accepted, to have in reserve in the event the employers gave notice.[9] This contained, as the most important change in the piece wage paragraph, a demand for a different basis of calculation equivalent to at least 50% of the hourly wage paid. With regard to this Ekman justified the executive board's proposal "which among other things had the intention of somewhat restraining the terror of the office which set the piece wages", with a choice of words worth noting. He admitted the right of the employers to carry out time and motion studies and to set up the necessary factory organization, but he noted at the same time that he was not insensitive to how the workers felt about this.[10]

As the Metal Trades Employers Association did not give notice of termination either the agreement was extended for a further year and thus was valid until 31 December 1928.

In the autumn of 1928 notice of termination of the agreement was issued by the union. A policy conference at the end of September to discuss the agreement revealed how powerful were the repercussions following the struggle in the spring against the legislation on collective agreements; the establishment of the Labour Court still exerted a hold on feelings in the Metal Workers Union. The representatives were almost entirely of one voice that the agreement should be terminated but there was a difference of opinion as to whether a new agreement should be made. Ekman threw the executive board's authority into the balance in favour of a new agreement and this view was confirmed by the policy conference.[11]

At the policy conference the discussion centred on piece wages to a much greater degree than earlier. Numerous speakers talked about the lack of reciprocity that followed the provision that the employer could cause even very slight changes in working methods or equipment to be carried out and thus justify immediate changes even in the fixed piece-wage schedules.[12] These views were met in the proposal for the collective agreement by an addition to clause 2 to the effect that piece wages could be altered because of alterations in the method or machines only to that degree which the alteration made it easier or more difficult to carry out the work.[13]

For the first time at such a policy conference of the metal workers, experiences of time and motion studies were presented by the representatives, not from the executive board. But it was only a single voice saying:

> An improvement has been introduced into the text of the new agreement, but the supervisors do everything possible to get the individual member to accept bad piece wages or to agree with a lower piece wage. To do this they use time-checkers and the

piece wage is set on the basis of the lowest time-check figure obtained. It is hoped that further changes could be taken in order to put a stop to this situation.[14]

But the speaker was not prepared to put this as a motion and this confrontation of actual time-study experiences with the text of the agreement thus left no direct traces. With regard to the other parts of the piece wage paragraph the new proposal for the collective agreement largely followed the proposal of the previous year. A demand for a basis of calculation amounting to 50% of the hourly rate paid reoccurred as did the clause on the bonus for dirty work.[15] These demands were considered to be especially important and the hope was expressed that they would be met to a greater degree than the purely wage demands.[16]

The basic positive attitude to rationalization, Ekman's work to a high degree, remained the unchanged basis of the union's demands. As an accompaniment to this should be regarded the paragraph about industrial democracy that was found once again in the executive board's proposal. The link between mutual understanding, contributory influence and rationalization, as already shown, was very much a reality at that particular time. At the conference on industrial peace a few months later and its continuation, where Fritjof Ekman played an important part, precisely the same note was struck.[17]

In the proposed collective agreement with which the Metal Trades Employers Association followed-up its counternotice for termination of the agreement no changes in the piece wage paragraph were requested. The negotiations were initiated under a neutral chairman.[18] It soon became clear that the workers' demands would only receive limited satisfaction. Half way through December the chairman submitted a proposal for settlement, which suggested prolonging the agreement with certain adjustments having no direct connection with the actual questions.[19]

The Metal Trades Employers Association accepted the offer with certain small reservations. The executive board of the Union also approved the proposal and distributed it for ballotting among the members. Despite some activity from communists in the union who wanted it to be turned down it was approved by a good majority. Just before Christmas in 1928 a new two-year agreement was signed.

The collective bargaining of 1928 introduced no particularly important changes in the piece wage paragraph of the metal trades' agreement. But rumbles of dissatisfaction were heard in the Metal Workers Union; the negative experiences of time and motion studies were largely responsible.

2. Conference 1930: the debate on rationalization

The next policy conference of the union on the collective negotiations was held in September 1930. The executive board had recommended that the agreement should not be terminated because of the uncertain economic conditions. Oscar Westerlund did not agree and joined the debate at the conference as an advocate for the opposition against the leadership. Economic conditions could get even worse during the years to come, he said, and the conditions set on foot by rationalization were becoming intolerable. In spite of all the arguments the conference accepted the executive board's line with an over $^2/_3$ majority.[20]

In the event that the employers gave notice of the agreement's termination, the conference adopted a draft agreement for the workers. The executive board's draft for a new piece wage paragraph differed on certain important points from the current one. A clause dealing with fixed piece wage schedules was deleted and it was suggested that all piece wages should have the same period of validity as the agreement. The demand was made for a basis of calculation of at least 50% of the hourly wage paid. Also new was the following addition to the clause dealing with how agreements on piece wages should be made:

> In those cases where piece wages are calculated by the calculation department or by means of time and motion studies, the worker shall be informed of the grounds for calculation and check the times.[21]

The demand for information on time and motion studies could be traced back to the link between contributory influence and rationalization. It was also analogous with what was laid down concerning bonus piece wages in the current agreement. Nevertheless, the fact that the demand for information was advanced meant that time and motion studies were no longer acknowledged by the workers to be the exclusive field of the employers but as an instrument for rationalization over which both parties had the right of insight.

In the discussions at the conference more voices than before were raised against rationalization and time and motion studies.[22] One deputy considered that the piece wage paragraph should include provisions about how effective time and lost time should be calculated in time and motion studies.[23] By means of alterations in clause 2 another, Ernst Ohlsson, from Gothenburg, wanted to ensure the right of negotiation of the local trade union organizations in matters concerning piece wages, and also that the employers' powers of altering these by reference to changes in method during the current period of agreement should be removed.

Ekman defended the executive board's proposal in terms that once again demonstrated his appreciation of rationalization. He commented on Ernst Ohlsson's demand about the negotiation rights of local unions as follows:

> The proposal made by Ohlsson concerning the alteration of clause 2 is impossible to utilize in practice. It would only achieve a mass of negotiations that would soon prevent the work being carried out . . .

And on the idea of attacking the employers' right to change piece wages:

The proposal ... shows that some people are not prepared to go along with the changes taking place in the process of production. Provisions that prevent a healthy development of the economy should not be included in the agreement ...[24]

In Ernst Ohlsson's reply he defended reciprocity in the agreement:

... Should the agreement prevent one party from obtaining alterations during the period covered by the agreement it should also place the same obstacles in the way of the other.[25]

The discussion between these two reveals the differing opinions about rationalization and the tactics to be used against the employers. These also constitute a factor in the more far-reaching confrontation between those who supported Ekman's line and the opposing group.

The policy conference approved in all important details the executive board's proposal for the agreement that was to be presented if the employers gave notice of termination. This did not occur, however, but the term of the current agreement was extended to the end of 1931.[26]

3. Collective negotiations 1931: The chairman's attempt

During the course of 1931 the conditions for collective negotiation changed because of the approaching and increasingly deeper slump that shook the world. In September 1930 the Metal Trades Employers Association had already discussed terminating the collective agreement in order to improve the cost position of Swedish industry but had refrained from doing this.[27] In the following spring it became increasingly clear that the Swedish Employers Confederation was preparing a unified, planned action to force down wages over the greater part of the Swedish labour market.[28]

The general council of the Metal Trades Employers Association at its meeting at the beginning of April had to consider a communication from the Swedish Employers Confederation concerning the possible termination of the metal trades agreement. During the ensuing discussion the managing director, Georg Styrman, painted a very black picture of the situation. The Swedish metal trades industry's pay was too high in relation to its competitors. Increases in domestic rates of pay were mainly due to piece wages. This fact, with the expectation that the workers' side would also continue to push its line to make all piece wages fixed, led to the conclusion that in the approaching collective negotiations the question of piece wages would be the central one. The board decided to prepare the question technically in the so-called apprentices committee, somewhat increased in size, before an answer was made to the Employers Confederation.[29]

The technical preparation resulted in a statement where the committee confirmed the importance of the Metal Trades Employers Association keeping firmly to its earlier line on piece wages:

namely, that fixed piece wages shall not be introduced unless it has been expressly stated from the concerned plant's management that the piece wages really shall be fixed. Therefore, proceeding from this, every such piece wage job, however many times it may be carried out, must in principle be considered to follow clause 3 of the agreement's paragraph 4.[30]

The committee made no proposal as to how the wages should be set but followed an argument in principle on this point. It was absolutely essential that wages should be reduced so that the minimum rates of pay as stated in the agreement must be decreased. But this was not sufficient:

The lowering concurrently of only hourly wages, however, is not really helpful as most of the work is carried out at piece wages. What is incontrovertibly most important is that the workshops receive lower piece wages. ... Thus the decrease may not be limited solely to such piece wages as have been set too high so that even following a reduction approximately unchanged earnings can be achieved. On the contrary the reduction must also be carried out in such cases where the piece wages are correctly set and where the reduction thus brings about a comparative decrease of the workers' income. Such a decrease should lie within the plants' contractual right, in so far as fixed piece wages do not exist. Probably, however, a more extensive general decrease does not appear possible of achievement through individual action at the workshops concerned.

The committee's report was presented at the end of April when the general council met once again. Georg Styrman thought that on the basis of what the committee had written together with the wage statistics assembled, contact should be made with the union side at the beginning of June in order to try to break down opposition to the lowering of piece wages and thus bring about an amicable settlement. Erik August Forsberg considered there was a risk in preliminary negotiations if these did not result in a very definite undertaking from the leadership of the union side, that it would not oppose a general reduction in the piece wages of, for example, 25%. On being asked by the chairman, Hugo Hammar, if a general agreement on piece wages did not carry with it a risk that they would be bound to regard all piece wages as fixed, Forsberg said he considered that there was no need for such a danger. Certainly it was a principle of the Metal Trades Employers Association not to discuss piece wages with the union organizations but in the existing position it was necessary and better to obtain an agreement centrally than to be compelled to negotiate with the local unions. Finally, said Forsberg, the risk wasn't so great if an agreement was obtained on a basis for the calculation of piece wages so long as the employers reserved the right to decide on the level of output. It may be noted here that at the time the general council was discussing a general lowering of piece wages the parties in the textile industry were going ahead with the reduction of piece wages, which was the main point in the agreement that followed the textile conflict (cf. pp. 163 et seq).

The general council decided to follow Styrman's recommendation for preliminary negotiations with the union and in accordance with this and following Forsberg's ideas, to reply to the questions from the Swedish Employers Confederation.[31]

In its reply to the Employers Confederation, the board stated that it agreed with the assessment that the situation with regard to costs was extremely serious and that the association was therefore prepared to go along with greater action for the lowering of wages *unless* the world's economic situation did not show a speedy and considerable improvement or *unless* an amicable settlement on a pay decrease was obtained with the trade union within its own field.[32]

The Metal Trades Employers Association immediately contacted the Metal Workers Union, but negotiations were not started until after the middle of June.[33] In spite of numerous meetings the negotiations were fruitless. The representatives of the Metal Workers Union in the opinion of the opposing side were not disinclined to accede to the demand for a decrease in piece wages on condition that the Metal Trades Employers Association was able to guarantee that the union's members would not be affected by any possible sympathetic lockout in connection with other conflicts on the labour market. However, the association was unable to provide any such guarantee without risking its relations and obligations to primarily the Swedish Employers Confederation.[34]

At its meeting at the middle of August 1931 the general council of the association discussed the situation and decided nevertheless to continue to try to obtain its aim without terminating the agreement.[35] If this did not succeed notice would be given. The aim was, as Forsberg stated the matter, not only to reduce piece wages where the workers with a higher tempo would compensate themselves, but "What the plants needed was less wages for a good piece wage job".

The council further discussed the lines of negotiation, especially the compensation the workers obviously required for their agreement on the wage question. The idea of offering a longer period of agreement, possibly with some kind of cost of living index provision, was put forward but did not lead to much discussion, as was the case with the idea of gaining approval for a lowering of piece wages by refraining from reductions in hourly wages. Once again the discussion showed that now piece wages were the key question.

Representatives of the Metal Workers Union had undoubtedly carried on their discussions proceeding from the proposal approved by the policy conference of 1930 on the agreement. A basic idea in this was that the difference between fixed and free piece wages should be wiped out so that all piece wages would be valid for the duration of the agreement and not subject to alteration unless methods were changed. Against this idea the Metal Trades Employers Association's principle of regarding all piece wages as free, unless expressly stated, had once again been confirmed before the start of the joint negotiations. The views were incompatible and as a solution of the problem the head of Svenska Kullagerfabriken (SKF), Uno Forsberg, suggested an idea that might make it possible to obtain a better agreement with the workers on the regulation of piece wages.

The minutes of the meeting state this as follows:

> With this purpose in view a suggestion should be made to the union of a commission headed by an engineer paid by the Metal Trades Employers Association together with two representatives appointed by the unions, with the task of investigating and reporting on demands for the lowering of piecework times in disputes about piece wages. If unanimity was not reached then the dispute should be dealt with in accordance with the steps laid down in the agreement for negotiation. The head of the commission should be assisted by competent time study engineers. Such a commission should be able to assist the whole of the Swedish metal trades industry to achieve greater competitiviness.

The suggestion is interesting and remarkable. It came from the head of a Swedish company that had one of the longest and most problem-free experience of time and motion studies and the setting of piece wages. The Metal Trades Employers Association had earlier on rejected all ideas of interference from the trade union in the setting of piece wages; and now Forsberg suggests a joint handling of disputes about piece wages with the idea that it was possible to establish a common norm, a common basis. The name of this basis was undoubtedly time and motion studies through which it should be possible to confirm the normal working time objectively for every operation. By using these studies a commission should be able to submit solutions of disputes about piece wages, which should contribute to industrial peace and the increased competitiveness of Swedish industry without internal strife. This line of thought portended a time and motion study council, which was to be conceived as a compromise solution out of the metal trades conflict almost 14 years later.

But in 1931 this went a little too far for most numbers of the general council of the Metal Trades Employers Association. Erik August Forsberg had various objections: the workers' representatives would be too strictly controlled by their organization; because of the large number of piece wages the commission would be overloaded with work, which would mean great expense for the association and lead to "consolidation of horse-bargaining on piece wages". Uno Forsberg's idea could not be permitted to conceal the demand for a decrease in wages. Neither was Georg Styrman enthusiastic:

> The Metal Trades Employers Association has maintained the mobility of piece wages and up till now succeeded; the workers have not dared to bring the matter before the Labour Court. The proposed commission thus meant a serious risk that piece wages might become an organizational matter, . . . The Metal Workers Union already had the idea of making piece wages an organizational matter; therefore far from encouraging such a development every suggestion on this should be fought . . .

Only Hugo Hammar expressed clear support for Uno Forsberg's idea according to the minutes.[36] It is quite certain however that the idea was never passed on to the Metal Workers Union's negotiating team as a compensation or compromise at the continued discussions at the end of August 1931. On the other hand the matter was discussed by the association's subcommittee appointed for negotiations with the workers. This resulted in a recommendation to the general council to set up "a time and motion study department of an advisory character that on request can assist the members in the regulation of piece wages and suchlike". As many of the members opposed the chairman's proposal, the question was shelved.[37]

At the meeting early in September the general council decided, somewhat contrary to Styrman's intentions, in the event of notice of termination of the agreement, only to put forward the demand for lower wages and otherwise recommend the prolongation of the agreement with the reservation, however, of being able to return with a completely revised proposal for agreement should the workers suggest formal changes in the agreement.[38]

When the Swedish currency went off the gold standard on 27 September 1931 a new situation was created that caused the Metal Trades Employers Association to reassess the decision on termination of the agreement. Following a discussion where Gustaf Söderlund, director of the Swedish Employers Confederation, was present and where loyalty to the other employers' associations played some part, the general council decided by 7 votes to 3 that notice of termination should be issued.[39] The employers' demands as formulated after the Metal Workers Union had presented its proposal mainly covered a cut in minimum wages of 10% and in piece wages of 18%.[40]

Preparations for the collective bargaining of 1931 in the Metal Workers Union have been described in rather great detail in earlier research.[41] At the meeting of the union's general council at the beginning of June, proposals from the employers' association about preliminary negotiations had already been received. In this situation Ekman recommended that the agreement should not be terminated by the workers' side and pointed out that the employers were certain to do so. Therefore the union should prepare for a conflict by raising funds by means of a levy. On the other hand Ekman saw no reason to call any policy conference about collective negotiations until it was possible to see how the situation was developing later during the autumn. In any case the agreement would be terminated and proposals for a new agreement had been available since the previous year. This decision aroused the latent conflicting opinions on the general council but the council followed Ekman's plans. It was decided, however, to wait with the levy until 1 October.

The division of opinion became still clearer at the general council's ordinary autumn meeting at the end of October. The employers and workers had already started negotiations under a neutral chairman. But the real negotiations did not get started until the middle of the following month. The employers took a stand with their somewhat modified demand for wage cuts whilst the workers' side had withdrawn to a position where they demanded unchanged benefits. A modiation compromise with reductions of 5 and 10% respectively was proposed and rejected and the negotiations were broken off. The leaders of the metal union then decided to call a policy conference on the agreement. The union's negotiating team had split into a majority led by the vice-chairman Oscar Westerlund, which did not approve of any decreases in pay, and a minority under Fritjof Ekman's leadership which wanted to play on differences of opinion in the employers' camp and seek a compromise with a 5–6% cut in piece wages and approval of certain other points of the workers. Thus the compensation idea was represented by Ekman, while the majority line thought that the negotiations would be wrecked and the conflict perhaps become a reality.

Many of the union representatives at the policy conference once again took up the negative experiences of application of the time and motion study system and of the reductions in piece wages it caused.[42] It was demanded that the interest of the workers' side should be satisfied in such studies. The demands were seized upon by Gunnar Andersson, a member of the general council, who at the same time made a suggestion about in which field the minority of the negotiating team thought compensation might be obtained:

> It seems to be wisest to follow the path of the minority, that is to say accept a smaller reduction against the substitution of more rigid wording in the piece wage provisions.[43]

The tone of the debate became quite acrimonious at times and most of the representatives sided with the majority. The minority argument was considered to contain "doubtful business speculations" and the idea of compensation was considered to be "horse-trading". Nevertheless the only decision arrived at by the policy conference was to leave it to the executive board to act should the employers carry out pay cuts at the turn of the year.[44]

A new round of negotiations between the parties was held at the middle of December, now under the leadership of a conciliation commission appointed by the government. A mediation proposal for reductions of 4 and 6% was accepted after some difficulties by the negotiating teams of the employers and employees, the board of the Foundry Workers' Union, which was now brought into the negociations, and the executive board of the Metal Workers Union. The proposal was distributed for ballotting among the members of the Metal Workers Union. Of the 27 617 participating, 16 441 voted against and 11 069 for the proposal which was thus rejected. The employers stated that reductions of 8 and 12% would be put into force on 18 January. In this position a meeting of the general council of the Metal Workers Union was called.

The discussions at this meeting covered a number of questions. The general council decided that certain workplaces would strike from 17 January, including 7 plants, and also discussed the economic situation and the amount of strike pay. But what is most interesting in the present context was that part of the discussion concerning the interpretation of the conciliation proposal on which had been based the latest poll among the members. In the circular, where the members were informed about the conciliation proposal, it was stated that no other reductions in piece wages were permitted during the validity of the agreement than the stated 6% unless changes in method were made at the same time. In other words the union's leaders maintained that although no formal changes in the paragraph relating to piece wages had been made, it had succeeded in gaining its point on the piece wage question. Most of the opposition did not accept this interpretation without a fuss. The different interpretations of the conciliation proposal were given a great deal of publicity in the press around the end of the year, which further increased the differences in the general council where Ekman's statement was criticized. Back summarizes Ekman's defence in the following way:

When Fritjof Ekman replied, he made certain statements regarding the piece wage problem that must be regarded as extremely optimistic, not to say sanguine, even though they were apparently made in good faith. According to Ekman nothing had been written in the circular to the branches that did not correspond with the contents of the mediation proposal. For throughout the negotiations the union's negotiating team had criticized the plants' application of the provisions about so-called temporary piece wages, and understanding had been obtained for this criticism. The latter should appear from the preamble to the mediation proposal (where it said that the current collective agreement should together with the supplement and the minutes be prolonged unchanged and thus retain its full validity concerning both pay and other provisions only with the adjustments and additions inserted in the mediation proposal), which would provide another meaning to minute No. 1 in the current agreement. In accordance with this, of course, the parties were said to be in agreement that the pay benefits, which were paid out in the special plants when the agreement came into force, could not be altered during the period of the agreement without voluntary agreement. Ekman must have meant that the preamble could be quoted as confirmation that the piece wages should be unchangeable during the period of the agreement. He said that he had been promised that he might call the conciliation commission's members as witnesses before the Labour Court to show the correctness of his point of view.[45]

In his thesis Back gives Ekman a slightly irresponsible trait and the reader gets a feeling that Ekman was prevaricating. This impression is increased when Back continues by characterizing Ekman's fiercest adversary, at the time, K. G. Taberman:

Long experience and a sure instinct told him that Ekman's point of view on the piece wage paragraph was badly substantiated.[46]

With this reservation about Ekman's reaction and Back's interpretation of it a link-up will now be made with the course of events, with a return later to the part played by Ekman and his grasp of the problem.

Notice of a strike by the workers was answered by the employers with notice of a general lockout. Faced by this threat the conciliating commission got the parties to return to the negotiating table; two days later a new conciliation proposal had been prepared where the lower limits of the reduced pay had been raised somewhat. This was approved by the negotiating teams and sent for polling among the members of the unions. Of the metal workers 17 991 voted for the proposal and 18 684 against while 135 ballot papers were blank. As the Metal Workers Union had all the time carried on parallel negotiations with the Iron and Steel Works Association and in intimate cooperation with the Union of Foundry Workers the executive board had the idea of totalling the votes from all members and thus was able to maintain that altogether 28 233 members had voted for and only 26 645 against. At the general council's meeting on 1 February, Ekman was thus able to show cause that the proposal should be signed. This was also the general council's decision by an overwhelming majority, 25 against 2, but not without bitterness and hard words from those opposed to the union's leadership. On 2 February a new one-year metal trades agreement was signed.

"On the surface the result was a resounding defeat for the workers" says Back and continues:

> In spite of assurances to the contrary the employers had obtained approximately the reduction they had calculated on. The original demands in the proposal of the Metal Trades Employers Association and the Iron and Steel Works Association had been provided with unusually large margins—margins that aroused opposition among their own people. The workers' side had not succeeded in obtaining any compensation provisions worth mentioning; the hopes raised in the context of a more favourable application of the piece wage paragraph in the metal trades agreement were to show themselves built on sand.[47]

Quite correctly the hopes of achievement were to be shown as weakly established. But it must be noted that at the time the agreement was signed, during the spring of 1932, it was still an open question. Ekman pushed an interpretation of the agreement that was not new in order to defend a bad agreement. Even the proposal for agreement approved by the policy conference of 1930 was based on the interpretation of the existing agreement that, during the negotiations in the autumn of 1931, Ekman said he had at least established with the conciliation commission. The conception mainly held by Fritjof Ekman is associated with his view of the possibilities of rationalization and the threat of time and motion studies. It was even to be of future importance for the position adopted by the Metal Workers Union and so for the relations between the parties in the metal trades industry.

4. The Labour Court 1932: guaranteed possibilities of pay

In the spring of 1931 Georg Styrman pointed out that the workers had not dared to carry the piece wage question to the Labour Court. However, the time had now arrived for Ekman to take this step also. The preliminary history of this lay a decade in the past.

In 1921 the employers gave notice of termination of the agreement for two reasons: to obtain reductions in wages and to get rid of the "freedom of movement" so highly valued by the union, that is to say the right during a current agreement to support new wage demands by strike action.[48] The demands were rejected by the workers. Attempts at conciliation were without effect and during 1922 the metal trades industry operated without an agreement. But renewed attempts at conciliation at the beginning of 1923 were to lead to an agreement approved by both parties. In this the employers were deprived of their freedom of action by means of a pay guarantee, which in the agreement received the form of the so-called minute 1. This obtained a simplified construction in the agreement of December 1923 and is found unchanged in the following agreements including that signed in March 1932:

> The parties are agreed that when the agreement comes into force pay benefits being paid at the special plants may not be changed during the period of the agreement without a voluntary agreement.

It was undisputed that this guarantee of the pay level during the validity of the agreement applied to hourly pay. The fixed piece wages were also included but it became the subject of disputes at the local and central levels as to whether the free piece wages in accordance with the agreement's paragraph 4 clause 3 were also guaranteed.[49]

The Metal Trades Employers Association held firmly to its line that if nothing else was expressly laid down, then every piece wage agreement was to be regarded as free and independent of the pay guarantee in minute 1. In the applicable disputes it had been possible to maintain this attitude quite well.[50]

The Metal Workers Union maintained on the other hand that even temporary piece wages should be guaranteed by this minute. It is in this context that the changes in the draft agreement of 1932 are to be regarded, where the differences between fixed and free piece wages are allowed to disappear and all piece wages are covered by the guarantee. Ekman went into the negotiations of 1931 with this offer. A proposal for changes in the text of the agreement was rejected but apparently Ekman received, or thought he had received, the support of the conciliation commission for his interpretation of the minute although this was not completely fully reflected in the text of the conciliation proposal, but only in an ambiguous phrase in the preamble.

Supported by this and the interpretation the union had made itself in its reply to the conciliation commission, in which it was stated that the union accepted a reduction of 6% but that greater cuts should be restored, Ekman considered that the time was now ripe to bring a suitable case before the Labour Court. In the Labour Court's case 1932/105 the question was therefore whether "the guarantee against the reduction of wage benefits only applied to hourly wages and piece wages in accordance with the fixed schedules or, in addition, other piece wages also, and in the latter case whether the guarantee applied to the wage level itself or the piece wage schedules applying at the commencement of the agreement period".[51]

In its suit submitted to the court the Metal Workers Union alleged that no kind of difference existed between fixed piece wages in accordance with clause 2 and temporary piece wages in accordance with clause 3, which was registered in accordance with clause 7. Therefore, and as the completely dominating number of piece wages appeared in accordance with clause 3, the union in the agreement of 1923 had assumed it to be obvious that the guarantee of the minute also applied to these temporary piece wages. The employers had, however, in practice increasingly ignored this and demanded a lowering of the temporary piece wages during the validity of the agreement. Therefore the workers' side in the agreement at the beginning of 1932 had in its reply to the conciliation commission especially stressed the importance that it should be confirmed that no reductions in addition to the general 6% might be made during the validity of the agreement that, the union also said, the conciliation commission had observed through its wording of the preamble to the final offer. In spite of this certain employers had demanded further reductions.

The Metal Trades Employers Association argued against this interpretation. The difference between fixed and temporary piece wages had to be maintained and reference of the Metal Workers' Union to the registration in accordance with clause 7 was irrelevant. At the agreement of 1923 the association had proceeded from the fact that the guarantee applied only to hourly pay because, in accordance with an earlier decision, in disputes about piece wages the work should be carried out on the basis of an hourly rate. The association had, however, agreed that special consideration should be given also to fixed piece wages. The agreement of 1932 changed nothing in the pertinent basis of the minute; it still did not apply to piece wages in accordance with clause 3. The association had certainly had its attention drawn by the conciliation commission to the demand for the locking of the temporary piece wages advanced by the opposing side. The association had, however, clearly opposed such an interpretation and it was therefore out of the question that the conciliation commission in its preamble had meant to give its support to the line taken by the workers.

The Labour Court was unanimous in its judgment. It dismissed the Metal Workers Union's reference to the preamble of the conciliators' proposal and based its decision on the interpretation of the minute of 1923. Nor was the court able to share the union's view that in practice no difference existed between permanent and temporary piece wages. This distinction must be maintained in order to avoid a pay system that was too rigid in a changing production. On the other hand it was clear that the temporary piece wages played an extremely large role in the metal trades industry. Therefore, contrary to the employers' view, the court considered it obvious that the pay guarantee of 1923 also applied to the *earnings* of temporary piece wages. The minute of 1932 had the same validity. This did not mean however, the court went on to say, that the individual *piece wage rates* had to be frozen for the period of validity of the agreement but what was important was that the *piece wage earnings* were not permitted to fall below the agreed, general reduction.

Naturally the judgment was, in its way, a reverse for the Metal Workers Union. Ekman had gambled on an interpretation of the minute which the court had not sanctioned. The path was no longer open for continued attempts to reach regulation and thus influence the setting of piece wages. On the other hand the judgment was also a reverse for the Metal Trades Employers Association. The pay guarantee had now been declared valid, if not to the individual piece wages, then to the total earnings from piece wages. This decision frustrated such plans as were aired in the spring of 1931 to arbitrarily lower the earnings-level of piece wages in the metal trades industry.

In his thesis Anders Victorin considers the Labour Court's decision 1932/105, strange because the judgment "is quite contrary to the grounds put forward by the parties". He presents the court's argument and sums up as follows:

> The Labour Court thus uses *the potential earnings* on piece wages as a basis for obtaining a norm, whilst the discussions between the parties had been about *piece wage rates*. The probability is that an approval of either party's claim would have led to consequences impossible for the other party to accept. As it was obvious that the Metal Workers Union's line of argument was wrong the choice lay between consenting to the Metal Trades Employers Association's application for rejection or to discover a third solution, which would maintain something of the balance between the parties. The solution it was possible to obtain by means of the arguments from the employers' side was singularly modern. It meant a collective regulation of piece wages at one and the same time on largely the same principles that later were to be introduced into the agreement through the regulation of the cash factor.[52] (*My italics*).

The court thus found an independent solution of the dispute by going beyond the arguments of the parties. It can surely be questioned whether the court in order to do this really needed the assistance of the Metal Trades Employers Association as Victorin suggests. It seems instead to be more fruitful to proceed from another opinion of Victorin and place the judgment in somewhat wider a context than that provided by the court's material. He writes:

> Minute 1 is very vaguely phrased and there have been a number of disputes about its correct meaning. Here the Labour Court has played a decisive role. In its interpretation the court has allowed itself to be guided by practical considerations and rather seen its task to be to achieve a ruling acceptable to both parties that could serve as the basis of a continued peaceful coexistence on the labour market.[53]

The court undoubtedly found itself in a tricky position. Case 1932/105 was so obviously a direct continuation of the negotiations during the collective negotiations earlier in the winter that the court's function in one sense was actually comparable with that of the conciliation commission. The court's speed, boldness and formal elegance under the chairmanship of Artur Lindhagen has been written about a great deal.[54] These characteristics also appear in judgement 105. There it was a case of creating the conditions for a future, better cooperation between the Metal Trades Employers Association and the Metal Workers Union by trying to eliminate the problem of piece wages as a source of dispute. The court could not declare that one of the parties was right and the other wrong, for this would have meant giving the antagonisms still greater scope. On the other hand it could not compromise in the way as a conciliation commission can. The court was compelled to present an independent new line acceptable to both parties. In one part the court supported the workers by agreeing that the temporary piece wages were covered by the pay guarantee, but supported the employers in another; the difference between fixed and temporary piece wages was upheld and the right to request changes in temporary piece wages existed for the whole term of the agreement. Finally, and this is what was new, the court interpreted the minute so that what was guaranteed was not the nominal piece wage rates, but the total of piece wages paid, that is to say the *earnings-level.*

Needless to say the court could not simply invent this if the parties were to accept it. But, with hindsight, what support could this line of thought count on from the parties? In the internal discussions of the Metal Trades Employers Association prior to the collective negotiations 1931/32, two alternatives appear, one rather radical and one more moderate. The former, with Georg Styrman and especially Erik August Forsberg as spokesman, wanted to obtain lower absolute wage costs for the metal trades industry and with this a real decreased level of pay for the workers. "What the plants needed was a lower wage for a good job of work", said Forsberg.[55] The more moderate line, carefully recommended by Hugo Hammar and Uno Forsberg, was to demand a lowering of the relative pay costs, that is to say that piece wage rates would be altered but only so that the workers with a high working tempo could maintain the pay they received. It should be added that it was in this connection that time and motion studies entered the picture, as method studies to make the higher intensity possible and as control studies to show the possibilities of keeping the pay-level constant. This justified the proposal put forward earlier by Uno Forsberg about a time and motion study council.[56] The Labour Court swept away the ground on which the harder line stood but was able in its interpretation to link up with the more gradual progress that Hammar and Uno Forsberg had advocated.

There were also differences of opinion on the trade union side, two dominant lines that had to be considered. There was Ekman's minority view, based on a positive view of rationalization and a policy of cooperation, and also the harder line of the majority with its main advocate in Oscar Westerlund, which was based on a harsh suspicion of rationalization and Mondism.

It seems reasonable to believe that the Labour Court's members, to the degree that in their decision they really permitted themselves to be guided by pragmatic or even tactical considerations, tried through their actions to avoid a complete defeat for the cause on which Ekman had staked his prestige. With regard to the principle validity of the minute, the Metal

Workers Union succeeded. On one point, the right to reconsider temporary piece wages, the court's decision was completely contrary to Ekman's. But in its motivation of this decision the court advanced considereration for continued rationalization, which was an argument associated with the idea of Ekman and his supporters. When it finally concerned the idea of a guaranteed possibility of maintaining an unchanged level of earnings with more intensive work, even this should have been comparatively easy to fit into Ekman's pattern of thought. He was no stranger to the idea that a more intensive working performance could be required of the workers. In the introduction to the discussion at the policy conference in December 1931 he defended his earlier work:

> Ever since I had any influence on dealing with such questions and since I became chairman of the union, the union has been able to see a steady increase in the metal workers' wages. Thus the cash pay has been increased by 16% plus a 14% increase in the money's purchaising power. The first increase is due to the union but not the second, which has depended on the decrease in living costs. *I am not unaware that the members have had to provide a greater achievement for the higher wage, but if, together with the other rationalization measures, this had not occurred, the metal workers would not have reached the point regarding their wages that they now have done.* In such a case we should now have seen greater unemployment figures. What has occurred has led to the country's metal workers now being found among the highest paid in Europe. Had a crisis not occurred it is possible that we should have been able to continue along this path.[57] (*My italics*).

Judgement 1932/105 thus implied, as I said, no great defeat for the Metal Workers Union, but contributed to the fact that the union was directed towards the alternative line of action that Ekman supported in the furtherance of rationalization. At the same time it was now up to the workers to obtain knowledge about, insight into and control of, the instruments for the setting of piece wages; initially time and motion studies, which were a condition for controlling the possibilities of earnings by piece wages.

To recapitulate, it is maintained here that the judgment handed down by the Labour Court in 1932, No. 105, linked up in a skillful manner with the existing ideas of both parties, and became for both sides a stimulation to the ideas of scientific rationalization, time and motion studies and an objective setting of piece wages, which were offered by the advocates of the rationalization movement. It is reasonable to assume that this development in the area of agreement made a powerful contribution to the great increase from 1932 in the employers' interest for time and motion studies, as shown by many observers. It also means that far from "putting a stop to the time and motion study frenzy", Ekman's policy was to lead to an increase in it. So it then became much more important to create guarantees of influence on such studies. The Labour Court was to return to this question in a new case whilst the collective agreement signed in 1932 was still valid.

5. The Labour Court 1934: piece wages of interest to the union

Case 1934/181 concerned the interpretation of the earlier judgment discussed above. In its application to the court the Metal Workers Union said that it was impossible for the workers to decide whether a new piece wage offer implied the possibilities of retaining the earnings-level. The employers had utilized this in order to force the piece wages below the level stated in the agreement. When the union had requested that this should be corrected the employers had also said that the minute did not raise any objections to the lowering of the individual piece wage so long as the average piece wage level *at the plant* did not go below that stated in the agreement. The union now considered that judgment 1932/105 must reasonably refer to the earnings-level *for the individual worker* and not for the plant as a whole. Therefore the union requested that in the fifteen individual cases put forward, the company should pay a retroactive remuneration to the workers who had been subjected to a reduction in earnings-level contrary to the terms of the agreement.

The Metal Trades Employers Association replied that the demand for an unchanged earnings-level in the individual piece wage was a new attempt to minimize the differences between fixed and temporary piece wages, which the Labour Court had undoubtedly laid down in its earlier judgment. Following this a number of arguments were put forward on the impossibility of fulfilling such a demand as the workers' side submitted. Piece wages were set under extremely different circumstances; what could appear to be work similar in character was often changed in some detail so that it was not a case of the alteration of old piece wages but a new setting of these. For natural reasons the earnings varied with the worker's temporary disposition and the piece wage often had the character of a common piece wage. Finally, the association thought that even if in certain cases it might be possible to show piece wages that had been set too low, the contrary could also be found:

> With the rational methods used nowadays for the study of the setting of piece wages cases were often found that were apparently wrongly set, as the worker reached high earnings in spite of an unreasonable waste of working time.[58]

In such circumstances, the employers said, it could not be considered otherwise than that judgment 1932/105 referred to the earnings-level at the plant as a whole.

The Labour Court completely rejected the arguments of the Metal Trades Employers Association and the company concerned: "By the responsibility stated in judgment 1932/105 for the employers to supervise the maintenance of the earnings-level is thus meant the earnings per unit of time of the special piece wage and not the combined earning of all temporary piece wages."

In view of the large of piece wages and the rapid technical development, the court continued, mistakes could of course occur. These could, however, not be blamed on the employer so long as he did his best to achieve a reasonable setting of piece wages in connection with the principles stated. Only if it was possible in a series of cases or in some individual cases of special principle importance to "detect a desire of the employer contrary to the obligations to which he was subjected to reduce the workers' earnings in temporary piece wages" could the employer then be considered to have violated the agreement and judgments.

With regard to the other matter in the Metal Workers Union's submission, which concerned retroactive pay to the wronged worker in the individual case, the court went against the union. Such a responsibility for payment would be much too unwieldy:

> The gain that the present system of piece wages carries with it, namely to facilitate a smooth adaptation to the conditions, could easily be lost. In actual fact the union's demand in this part is identical with that demand disallowed by the Labour Court in the earlier case.

Then how was liability to be claimed from a negligent employer if the demand for payment to the injured party was not approved? On this point the court once again, in my opinion, made a new construction, which went beyond the parties' arguments in the actual case but which became of principal importance in future development. The court wrote:

> The question then is in what manner rectification shall be obtained where the employer disregards the responsibility put upon him to supervise that in the setting of new piece wages the earnings shall if possible be maintained. In the opinion of the Labour Court this question cannot be answered in more than one way. The employer's responsibility referred to must be regarded as a responsibility not in relation to the individual workers but to the union as the contracting organization on the workers' side.

With this the court thus rejected the demand for payment to individuals, with reference to the consideration of the flexibility for adaptation of trade and industry, which was an argument the leadership of the Metal Workers Union found difficult to counter. On the other hand the court approved the idea that the pay guarantee applied to the individual piece wage, which was again for the workers' side. The view that the guarantee was an obligation between the employers and the union, finally, and that in a breach of the guarantee *the union* could take action, was in fact support for the line pressed by the workers' side. This made the individual piece wage into a matter for the union organization, which the Metal Trades Employers Association had consistently opposed. Thus judgment 1934/181 was a greater defeat for the employers' side.

6. Summary and remarks

This chapter has critically examined how the discussions on the collective agreement within the Metal Workers Union towards the end of the 1920s were marked by disagreements between a line positive to rationalization and one that adopted a negative attitude. The union's chairman, Fritjof Ekman, was to personify the positive point of view. In this context it is worth quoting Per Erik Back's assessment of him:

> To a certain extent the sometimes negative conception of certain members with regard to Ekman's policy depended upon the fact that he as chairman—perhaps more strongly than the trade union movement's leadership in general usually did—fostered the program that in the short term was likely to create difficulties for the workers but in the long term should strengthen their position. The workers' position should become better because the situation of the Swedish metal trades industry became better—this Ekman believed and he propagated with energy and consistently for a positive attitude to the entrepreneurs' efforts to rationalize and modernize.[59]

During the period with which we are concerned here, time and motion studies arose as a source of irritation. At the policy conferences to discuss the agreements in 1928, 1930 and 1931, the critical voices grew in number and volume. However, these studies were not regarded by the leaders of the union as any real threat but rather one of the employers' legitimate tools in the rational setting of piece wages. On the other hand, to the extent that time and motion studies were used to lower piece wages, the results were something for the union to deal with.

With regard to the wage guarantee for piece wage workers, Ekman followed his own line. Since 1923 the agreement had contained a pay guarantee in the shape of minute 1. Its validity with regard to hourly wages and fixed piece wages was not questioned, but there were different ideas about whether it also applied to temporary piece wages. Ekman's tactic to get this pay guarantee approved all along the line was, from 1930, to try to eliminate the agreement's classification into fixed and temporary piece wages by removing the special provisions on fixed piece wages and to maintain that all piece wages found in the metal trades' piece wage registers were in practice fixed and should be treated as such. A de facto admission of this equality between different types of piece wages, said Ekman, had been obtained during the conciliation negotiations in the autumn of 1931. On the basis of this the Metal Workers Union went to the Labour Court in 1932 to get this interpretation sanctioned. The result was the opposite: Ekman's policy suffered a defeat.

It is possible that Ekman, because of the internal criticism, was compelled at the wrong time to apply for the adjudication of his line to the Labour Court, and the foundation was not wholly convincing. But it should be clearly understood that Ekman followed a consistent policy in the piece wage problem in which the supporting element was a combination of minute 1 and the removal of the categorization of piece wages.

But where Ekman lost the opposing side did not win a great deal. The Labour Court launched a line of its own in its interpretation of the adaptation of minute 1 to the requirements of modern industry for changeability. The pay guarantee was declared to apply to earnings' potentials on piece wages also. This meant a disengagement from the nominal piece wage rates that stimulated, in fact presumed, a more sophisticated method for the calculation of piece wages—and for the control of potential earnings. These methods were available in the shape of time and motion studies. The Labour Court's decision thus came to influence not only the formation of norms in the area of the metal trades agreement but also the palpable reality of the metal trades industry's daily life for decades to come. In its interpretation in 1934 of the earlier judgment the Labour Court made a further important ruling. Piece wage questions were declared to be a matter for the organizations rather than for the individual worker: it could never be decided in an individual case whether the spirit of the agreement was being followed by the employers.

CHAPTER 4

The Build up

1. Collective negotiations 1934–35

The agreement of March 1932 did not solve the problems of piecewages for the Metal Workers Union. On the contrary, because of the Labour Court's interpretation the agreement became a further incitement for the employers to introduce time and motion studies to increase working tempo or to discover changes in methods that made reductions in piecewages possible. Dissatisfaction about the piecewage problem also flowered within the union.

Towards the end of August 1932, as mentioned earlier, an ordinary Congress of the union was held. In the discharge of liability debate a powerful attack was made on the leadership's way of dealing with the collective negotiations during the previous winter. Not least the unsuccessful attempt to interpret minute 1, was subjected to criticism. It should be observed, however, that Gunnar Andersson, who later the same year was to be elected chairman, wanted to absolve the executive board on this point and thought that there had been justification for the optimism criticized.[1]

The policy conference held in connection with Congress once again revealed a split between the radicals from Gothenburg, who in spite of the sombre economic situation wanted to terminate the agreement, and what was now the large majority which found that the slump made every assault meaningless. Those who demanded that notice should be given mainly advanced the provisions on piecewages as their motivation for this.[2]

On the other side the employers association discussed termination of the agreement at the meeting of the general council in September. The Employers Confederation had recommended a general action for cuts in wages, which was made possible as unusually many collective agreements were running out and could be terminated. For various reasons, however, many of the general council's members formed the opinion that the metal trades agreement should be extended if nothing unforeseen occurred. As a basis for this decision lay, with other factors, the situation that the Swedish pay level had now been reduced in relation to other countries and that the economic prognosis did not point downwards unambiguously.[3]

Nor was the agreement terminated by either of the parties in 1933. In the Metal Workers Union not even a policy conference was called to discuss a possible round of collective negotiations.[4] The Metal Trades Employers Association's general council certainly discussed once again the idea of an offensive for the reduction of wages at the last minute but most of the members adopted a wait-and-see attitude. The general council kept to its assessment that to give notice to terminate the agreement was not suitable for the metal trades industry, but if the Employers Confederation wanted to carry out a general action to support wage negotiations within domestic industry, the association would be prepared to consider taking part in this.[5]

The passivity on the part of the workers during 1932 and 1933 is obvious in view of the economic situation. The discussions within the Employers Confederation and its decision to abstain from new threatening actions must be considered to imply that it was so satisfied with such things as the interpretation of the piece wage paragraph that uncertainty about development of the economic situation was sufficient to restrain it from taking the offensive.

In the autumn of 1934 the economic position had so changed that the roles, in one way, were reversed. When the general council of the Metal Trades Employers Association discussed the position with regard to the agreement its chairman, Georg Styrman, was able to assume that nobody in the association intended to request termination. But rumours were in circulation that movements within the Metal Workers Union were working for an offensive. The council decided that it would only issue a counternotice should the union give notice. It also discussed the tactics to be used in possible collective negotiations and was unanimous that at an early stage it should state its intentions to institute a lock-out if no agreement had been reached by the end of the year.[6]

The Metal Workers Union had its usual policy conference in September. It came as a surprise when the executive board's recommendation to try to prolong the agreement was announced. The executive board said that certainly the economic development looked promising and certainly there were a number of points in the agreement that the union had cause to be dissatisfied with, especially holiday and piece wage problems, but uncertainty in the economic and political situation around the world urged some degrees of caution. It was to be shown that the executive board's opinion was not shared by the majority of the representatives and so the debate will be examined in comparatively careful detail.

Gunnar Andersson, who had now become the union's chairman and therefore opened the discussion in the customary way, also raised the piece wage question.[7] There was a certain degree of bitterness in his speech as Back observes, not least against the judgment handed down by the Labour Court in 1932.[8]

But there was more intense criticism on the same point from the more radical wing, put forward by the Gothenburgers who wanted to use the court's interpretation as the reason for issuing notice of termination.[9] Oscar Westerlund was not prepared to go so far but his description of the situation did not mince words either when he spoke about the effects of the Labour Court's judgment of 1932:

> ... means, that the workers are surrendered unconditionally to the employers, as nowadays time and motion studies are being introduced everywhere and piece wages altered accordingly. The workers have to work harder and carry out more operations than earlier—but in the best case receive the same pay as before. This puts the conditions in the workshops in a new light and the workers can hardly continue to work in the conditions now existing. What is most important is that the agreement gets a correct paragraph about piece wages, by which is meant that the piece wage price cannot be altered during the period of the agreement if lesser mechanical changes are carried out ... My idea is thus that notice of termination of the agreement should not be issued this year but that we should wait still another year and see how the economic situation develops. Then the members will have time to stabilize their economy, especially those who have been unemployed for a longer time. The day we are able to arrange the piece wage question it will be no ordinary conflict but a great conflict ...[10]

Gunnar Andersson's reply to the attack betrayed the agony of the responsible pragmaticist.

> The union's most difficult task is to be able to arrange the question of piece wages satisfactorily within the industry as it has shown itself to be impossible to obtain a real grasp of this question. If somebody could construct a satisfactorily formulated agreement that we could later get our adversary to accept, much would be gained. Whatever phraseology is used in the agreement we must always count on the circumstance that the two sides have not agreed about a piece wage, and what's to happen then? ...[11]

The most far-reaching demand, however, that piece wages should not be altered at all even when mechanical changes were carried out, was turned down with reasons that recalled Ekman:

> Should such a provision be introduced it would be the equivalent of saying that development within the industry should cease, and presumably the idea is not that the trade union organization shall cooperate in this?[12]

The ball from Andersson's court was returned by K. A. Åberg from Uppsala, who spoke from his experience of Nyman's workshops there. He considered that the question of time and motion studies could well be arranged in agreements that contained provisions about hourly pay and supplementary percentages.

The discussion resulted in the conference, by 30 votes to 24, deciding to give notice of termination of the agreement, contrary to the executive board's recommendation. Following this there were discussions about the board's proposal for a new agreement. The important proposals for changes were in line with the policy carried out earlier. The clause relating to fixed piece wage schedules should be deleted and replaced by a provision that all piece wages agreed should apply during the validity of the agreement provided no mechanical changes were made. All piece wages should be set so that they gave at least time and a half. Where time and motion studies were concerned the basis for calculation and method should be laid down by agreement, that is to say the workers demanded joint determination in time and motion studies. With smaller additions and changes the executive board's proposal was adopted.

In due course notice was given to terminate the agreement and the Metal Trades Employers Association presented a counterbid with decreased amounts for minimum pay and other changes disadvantageous to the worker's side.[13] The status quo was claimed for the piece wage paragraph.[14]

The following collective negotiations were very tough and broke down already after an initial round and a conciliation commission was appointed. No result was achieved by this and the employers gave notice of a lockout from the new year. During the week before Christmas new negotiations were set on foot and a new conciliation proposal was presented. This contained no changes in the most important questions, the holiday and piece wage paragraphs. In the workers' negotiating team it was decided to concentrate efforts on the holiday paragraph as compact opposition of the employers was expected on the piece wage question and there was uncertainty about what real value would be obtained from formal changes in the piece wage paragraph.[15] The executive board approved this line, especially as the Labour Court's decision in case 1934/181 seemed to supply some of the security sought in the piece wage question. The position taken up by the negotiating team was wholly in line with the pessimistic assessments of the chances of success already made by the leadership at the policy conference on the agreement. Back, who only shows the excecutive board's later attitude, appears to have overestimated the importance of the Labour Court's decision for the continued tactics in the negotiation situation, which conceals the continuity and actions of the executive board.[16]

Concentration on the holiday provisions produced results. In a new conciliation proposal, which was approved by both parties, an increase in holidays by one day was gained for the workers and also an upward adjustment of piece wage compensation. The retreat by the workers on the piece wage front was masked by their request to the Metal Trades Employers Association for a declaration that workers' benefits would be left untouched during the validity of the agreement and that the piece wage paragraph so applied that even the individual member would obtain protection from it. The Metal Trades Employers Association rejected with vigour this demand outside the agreement. A new agreement to last for one year was agreed on at the beginning of January. The negotiators of the Metal Workers' Union stated that the demand for the suggested declaration was withdrawn as the employers' statements during the discussions had been understood as declarations in the required direction.[17] The agreement reached at the beginning of 1935 thus meant no changes in the piece wage paragraph but the relation between the parties had, however, shifted in favour of the workers through the decision of the Labour Court. In addition the trade situation improved and the national economy became stronger. Was the year that lay ahead to bring the great trial of strength about piece wages?

2. Collective negotiations 1935–36

Even the statement by the union's negotiators in conjunction with the agreement reached at the beginning of January 1935 was regarded by its opponents as noise of battle, with the threat of renewed notice of termination. The actual basis of this threat was rejected by the negotiating team of the Metal Trades Employers Association—it applied of course to undertakings to modify the application of the piece wage paragraph—but nevertheless caused the negotiating team to propose to the general council an increase in the preparations on the threatened front:

> However, within the Metal Trades Employers Association's negotiating team the idea had arisen that the association should employ an engineer with experience in time and motion studies and modern methods of piece wage setting who could be used in disputes about piece wages that might arise to investigate carefully the piece wage conditions at the workshop in question and prior to possible negotiations in the dispute provide an objective report on the way the employer in question dealt with piece wages. It was not intended to compete with existing consultants: the man should only act on the commission of the Metal Trades Employers Association.

Thus Uno Forsberg's proposal from 1931 was presented in a modified form. No doubt not least the Labour Court's reactions contributed to the fact that the proposal now claimed a majority in the general council of the Metal Trades Employers Association.[18] From August 1935 a 28-year old university-trained engineer, Gösta Ingelman, was appointed as time and motion study expert.[19]

The ordinary Congress of the Metal Workers Union was held in September 1935 and also functioned as a policy conference. Among the motions placed before Congress was one from Sandviken which dealt with the consequences of time and motion studies. This resulted in a proposal to the general council:

> in future negotiations on the agreement to put forward demands for a provision in the agreement that protects the workers from unjust treatment through time and motion studies.[20]

In its comments the council proposed that the motion should be referred to the policy conference for working out in detail. In Congress, however, the agenda was so ordered that questions concerning the national agreement were dealt with first. The motion, supported by many speakers, could therefore later be met by Gunnar Andersson with a statement that the important demands had already been satisfactorily met in the union's draft for the agreement. Congress accepted the general council's statement.[21]

Treatment of questions concerning the agreement was initiated by Gunnar Andersson, who developed the reasons that had led the executive board to recommend that notice to terminate the agreement should not be given. With regard to the piece wage paragraph the employers were certainly just as much against every change of its formal content as the previous year, for since then the decision of the Labour Court had been given and its application had become more pliable.[22]

The executive board's defensive attitude was attacked from many quarters, especially with regard to the piece wage situation, and when the vote was taken as many as 109 delegates to 61 voted for termination of the agreement.[23]

In the following negotiations both the Metal Workers Union and the Metal Trades Employers Association presented proposals that in the piece wage paragraph were worded exactly as in the previous year's; for the employers they were worded exactly as in the current agreement.[24]

Earlier in the year Gunnar Andersson appears to have been cautiously optimistic about the possibilities of obtaining at least a more satifactorily application of the metal trades agreement, especially in the piece wage paragraph, feeling it should thereby be possible to establish a more relaxed relationship with the Metal Trades Employers Association.[25] Antagonism between the parties, which as was said in many quarters also had a personal basis, had at the time of the negotiations reached the point where it was thought fit to link the negotiations directly to those being carried out between the Metal Workers Union and the Swedish Iron and Steel Workers Association (Järnbruksförbundet) and to use the same conciliation commission.[26]

Externally the progress of the negotiations resembled those of 1931/2. Many rounds of negotiations together with the conciliation commission were without result and were followed by notification of a general lockout by the Metal Trades Employers Association, starting on 7 January. This threat forced the parties to resume negotiations and after three days' intensive and long-drawn-out negotiations, it was possible to issue a draft agreement prepared by the negotiating teams for balloting the members of the Metal Workers Union.

This proposal did not meet the workers' demands about changes in the piece wage paragraph. What applied essentially here of course, was that the agreed piece wages should be valid for the whole period of the agreement, that piece wages should yield at least time and a half and that the basis of calculation and the time and motion study procedure should be confirmed through agreements and not be a matter solely for the employers. On the last-named points the conciliation proposal provided some progress and prescribed the right of the worker involved to information about the methods used. Furthermore, in certain cases piece wages arrived at through time and motion studies could be added to the fixed piece wage schedule. Also there was a proposed addition to the minutes to the effect that negotiations on the fixed schedules should be arranged at which central negotiations could also occur.[27]

On this occasion also the result of the ballot on the proposed agreement was extremely close. There were 20,536 against and 19,317 for, with a resultant majority of only 1,219 votes for rejection. On the other hand, in the iron and steel works area a majority of 2,611 votes was obtained for it. The result led t a lengthy discussion on principle in the general council—whether it was possible, as in 1932, to combine the votes for the members in different trades (cf. p. 71) and so obtain a majority for the ayes. This resulted in the general council's support for the chairman's line. The negotiating team was to be given the task to conclude a new agreement on the basis it was quite certain that a conflict would be fruitless. The negative result of the ballot on the metal trades agreement was, however, to be utilized in continued negotiations in order to compel changes in the final conciliation draft.[28]

So a number of demands for alterations were advanced in the continued negotiations before the conciliation commission: a one-year agreement (the proposal was for two years) and changes regarding shift work and hourly pay bonuses. Detailed suggestions for alterations were also presented in the provisions of the conciliation proposal for the piece wage paragraph.

It turned out to be impossible to achieve any concessions on the validity of the agreement although gains were made in some of the other items. But not in the piece wage problem to which most of the final phase of the negotiations was devoted. The negotiators for the Metal Workers Union reported:

> but it was completely impossible to obtain any result as the employers seemed to be afraid that in this way the organization would be involved in the setting of piece wages in accordance with paragraph 4, clause 3. The director of the Swedish Employers Confederation, Söderlund, advised us to arrange this question by sending out rules of conduct to the members implying a prohibition against consenting to the fixing of piece wages without first obtaining the approval of the local union concerned.[29]

The phrasing shows that the stumbling-block in the final discussions was the demand for agreement between the workers and employers about the time and motion studies in clause 3. Both when it applies to the addition of other piece wages to the fixed schedules and when it applies to the right of information on time and motion studies in clauses 3 and 4 in the final agreement, the terms "worker in question" and "worker concerned" are stressed, that is to say the organization is excluded.

A two-year agreement was signed on 21 January 1936 with changes in the piece wage paragraph, which was still far removed from the demands expressed in the Metal Workers Union and which their negotiating team had pressed to a certain extent in the negotiations.[30] The result also caused some criticism in the union, especially from Gothenburg.[31] Back's summation of the result says:

The result of the wage negotiations was undoubtedly very limited. In everything of importance its character resembles that which marks the movements of the long defensive period. The Metal Trades Employers Association's attitude was unmodified and expectations of the union's leaders were modest to start with. "Purely employers' arguments" were advanced by the leadership to an astonishing degree.[32]

3. Policy formation of the Swedish Metal Trades Employers Association

Nevertheless the result of the negotiations in January 1936 should not be underestimated from the workers' point of view. By virtue of the provision, that the worker concerned was entitled to be informed about the grounds of time and motion studies, an initial and important breach had been made in the employers' opposition to surrendering their uninhibited power over the setting of piece wages. Time and motion studies had become a question for agreement, a legitimate sphere of interest for both parties, and it was now necessary for the Metal Trades Employers Association to take up the question in a more resolute manner.

The association was better equipped for this purpose since the summer of 1935 than it had been earlier as Gösta Ingelman had been appointed (cf. p. 85). The agreement of January led immediately in the following month to a fundamental discussion in the association's technical committee about the grounds for the calculation of piece wages. A memorandum on the subject had been circulated prior to the meeting. In addition the whole conglomeration of problems had been previously discussed by the Royal Academy of Engineering Sciences. The academy's committee on time and motion studies had arranged discussions and lectures on the subject on the 25–26 February. A German rationalization expert had been present, Dr. Peiseler, who spoke about "Time or money piece wages".[33] The technical committee's meeting the following day was to revolve around these key words.[34]

In order to understand the discussion one postulate must be presumed, namely the Metal Trades Employers Association's principle view that the calculation of piece wages was something for the employers alone. If the worker was not prepared to accept a piece wage offered he was entitled to hourly pay, but he had nothing to do with how the employer calculated the piece wage. The worker was offered a price and he was entitled to negotiate about that with the employer until a price was agreed: a money piece wage.

The advantage of time and motion studies as a system for setting piece wages was now that the time required for a job could be measured in accordance with fixed norms. In a piece wage equation the time measured could be placed in direct relation to a certain earnings-level and disputes about piece wages should, in theory, vanish. But when the "normal earnings-level"—the cash factor—was locked through an agreement about the piece wage excess or through a direct agreement on the cash factor or through an earnings-level guarantee, the character of the piece wage system was altered. In, for example, a central agreement about the cash factor the earning were no longer a variable in the equation. It was no longer possible to bargain about the price. Instead it was about the other factor, time, that the tug-of-war between the workers and employers was to be fought: time piece wages. When the system was supported by objectively agreed methods for time measurement it was more agreeable for the workers as more permanent and secure. Gains made at central negotiations about the money factor could not be conjured away so easily at the local individual piece wage negotiations. Every step towards locking the cash factor thus came to be regarded as a gain for the worker's side. If the system with time piece wages was fully constructed it was up to the employers to try to retain some degree of freedom of action in the matter of deciding the time, that is to say, the amount of work. It was also to appear that the employers' interest in the so-called lost time and performance index factors increased as time went on.

One step along the path towards the locking of the money factors applied in practice was the Labour Court's decision of 1932 that guaranteed the earnings-level. But still the workers had no real insight into how the piece wages were set. The cash factors used by companies for their calculations were secret. But when following the 1936 agreement the employers became liable to show on request the grounds for time and motion studies, including the principle for the relation of the time measured to the cash factor, it was easy for the workers to calculate which cash factor was used in each individual case.

In the memorandum to the meeting of the employers' technical committee on 27 February it was also stated that it was presumably difficult to keep secret which cash factor had been used. It should therefore be just as well to recommend that they were reported openly instead. On the other hand it would be unfortunate if the cash factors were to be fixed at definite values. Instead they should be stated within certain limits. In this way the time and motion study man should be able to use varying values, among other things with reference to the possibilities of building up the piece wages. In the setting of piece wages, thought the author of the memorandum, the time and motion study man should also take note of the workers' earlier earnings.[35]

The discussion showed the differences of opinion existing among the members of the committee. The association's vice managing director Gösta Göransson, as also Gösta Ingelman, argued for the line of the memorandum and thought that a completely open cash factor was required in order to be able to use fully the advantages of time and motion studies carried out correctly. They were supported by some of the committee whilst other members were in principle against this interference in individual concerns of management. The committee's chairman, Erik August Forsberg, was mainly in favour of the memorandum's line, but wanted to remove the text prescribing that regard should be paid to the worker's earlier earnings in the setting of piece wages. This also became the committee's decision. The committee stated that the memorandum with the mentioned alteration, "should be able to serve as a guide for the Metal Trades Employers Association's office in questions from the members about problems in this field. This presumes that in the answers given the greatest attention is paid to the different conditions at the company concerned." There was no intention of distributing the memorandum as a statement of policy. With this the Metal Trades Employers Association had formulated a principle line in the piece wage question in so far as time and motion studies were concerned. This line meant an affirmation of the possibilities offered by time and motion studies whilst being aware that this implied it was necessary to waive certain earlier rights of the employers. But this attitude was not unanimous and the association was still not ripe for any policy carried out centrally. Still less were the employers prepared to give the workers the right of negotiation on money factors. The discussions so far had only covered the right to information.

4. Collective negotiations 1937: a new policy

In the autumn of 1937 when the metal trades agreement could be terminated again the conditions, as Back points out, had altered in important ways. The economic situation was stable at a high level. There was a new leadership of the Trade Union Confederation, where August Lindberg was chairman and Gunnar Andersson vice chairman, which introduced a more active and concurrently a more peaceful union line. The negotiations on a general agreement were about to begin and a policy of compromise was once again being advocated. Combined with the Trade Union Confederation's interest in compromise there was a strongly positive evaluation of and care about the efficiency of trade, echoed in statements from the leadership of the Metal Workers Union.

At a general conference about the new agreements in August the Trade Union Confederation recommended preliminary negotiations and agreement across the board.[36] Disturbance of the economic situation by collective negotiations was to be avoided so far as this was possible, which was also a way of achieving a better grasp by the confederation's leadership over the unions' actions.[37] This control with regard to the Metal Workers Union was carried out through the industrious presence of Gunnar Andersson at the union's policy conference.[38]

It was not difficult for the Metal Workers Union to understand the compromise signals in relation to the Iron and Steel Works Association. The importance of Gustaf Rudenstam as managing director of this association has been vouched for by the historians of both the Metal Workers Union and the employers association.[39] In a corresponding although opposite manner, the Metal Trades Employers Association's director Georg Styrman was regarded by the workers as the personification of the employers' obstinacy. The bad relations between the parties caused the chairman of the Metal Workers Union, Oscar Westerlund, to adopt the tactics of trying to split the employers. Preliminary negotiations were started with the Iron and Steel Works Association without notice of termination of the agreement being given and, following certain difficulties, carried to an agreement on 11 November. However, notice was issued to the union's other counterpart, the Metal Trades Employers Association. It is uncertain whether this depended on misunderstandings between the parties or a lack of will.[40]

The Metal Workers Union's policy conference to discuss the new agreement with the employers was held 17–18 September. The executive board recommended notice of termination, but with comparatively limited demands where the main emphasis was on pay increases, the definition of a skilled worker and the holiday paragraph.[41] Thus the questions of piece wages and time and motion study did not come into the problem areas given priority. This assessment was sharply criticized by many speakers who with various degrees of intensity emphasized the importance of changes on this point.[42] But none of them seemed to have any real proposals for solutions with the exception of some who considered that special plant representatives should be appointed with the main task of checking time and motion studies.[43]

Oscar Westerlund, the new chairman of the union, was not convinced that this was a practicable path to follow:

> Which of you would like to be such a trusty aid? It is important we don't forget that quite frequently, especially with regard to questions about piece wages, he may be compelled to go against the workers when he sees they are wrong . . .[44]

The chairman continued by stating an attitude to the piece wage problem and to time and motion studies as a system, an attitude, that bore witness to the way in which his radical point of view had been modified:

> Is there anybody who in this matter is able to present a really complete idea as to how the piece wage problem is to be solved? It is not enough just to consider what suits a group of workers as this may be damaging to another, and what is most important is that a system is obtained that is suitable for both workers and industry.
>
> In the discussion earlier today the system of time and motion studies at the workplace has played an outstanding role. But at the same time we must remember that we are living in a time of rationalization and that these studies have become a science, which to those who are engaged in them retain their value without reference to psychological conditions. In my belief the time will come when it is shown that when the time and motion frenzy has been allowed to continue for a period there will be a setback. The cases where there will not be a reverse are those where the work recurs and is similar, in which case there will still be a use for the system. At those places where they have temporary piece wages, this must be abandoned. I am not completely convinced that we should fight for this matter, for it is much too new and complicated. There are examples of where the worker who has been opposed to the

time and motion study system has later found it to be good after using it for a period. It depends on how the workers are able to maintain their rights ... If a decent agreement is reached with management and guarantees are given that management understands that this is not only a scientific system for calculating time but that also psychological considerations must be introduced, then we have made some progress.[45]

On some points the board's proposal for the piece wage paragraph of 1937 differed radically from the line that the union had pursued in the earlier collective negotiations. As late as the 1935 negotiations the workers had tried to increase security in piece wage setting by erasing the difference between fixed and temporary piece wages and by giving all such wages unchanged validity for the period covered by the agreement. It became increasingly impossible to push this line after the Labour Court's judgment 1934/181 had once again confirmed that the importance of the difference between types of piece wages being maintained, and since the agreement of January 1936 had created a stimulus for the creation of fixed schedules.

The executive board's proposal for the agreement of 1937 confirmed that fixed schedules for piece wages should be set up where this was possible (clause 2) and that, in the exceptional cases where this was not possible, the parties should agree about the temporary piece wage before the job started. Should time and motion studies be used the grounds for calculation and the method should also be the subject of agreement (clause 3). The piece wage should be time and a half for the worker concerned. All piece wages should be increased by 15% and be valid for the whole period of the agreement (clause 4).

What was most important in the new proposal was the emphasis on the significance of the fixed schedules and the attempt by means of the expression "in exceptional cases" to circumscribe the employers' possibility of utilizing a temporary piece wage in accordance with paragraph 4, clause 3. In the discussions that followed the proposal was wholly approved by the conference, the importance of really obtaining fixed schedules being stressed. In addition to certain separate and more radical proposals, which were turned down with reference to the fact that time and motion studies had come to stay, it was really only the demand for trade union representatives to be present at time and motion studies that was not contained in the board's proposal.

Negotiations got under way at the middle of October but so little progress was made that they broke down a month later and a conciliation commission was appointed. Behind what was in fact a delaying action were the destructive tactics of the Metal Workers Union and when negotiations were resumed in mid-December they were carried on under the powerful impression given by the agreement made by the iron and steel works. The iron and steel works agreement of 1937 has been considered to have had especially far-reaching scope as the iron and steel works took over the role of the metal trades plants as the model for wage negotiations in a much wider sector of the labour market. In addition, through the interplay of various circumstances, this agreement was to produce effects over an unusually long period.[46]

Just before Christmas the conciliation commission managed to get the parties to agree on an offer that was put to the ballot of the union members during the Christmas holidays. The majority for an agreement was small but indisputable and the agreement, a 2-year one, was signed on 13 January 1938.

Among the workers there was satisfaction with the new agreement as a number of gains had been made. Minimum and hourly rates of pay had been increased by 7 öre (1 Swedish crown = 100 öre) and all piece wages by 7.5 %. With regard to shift and overtime work and holidays, work in another district and industrial welfare, advances were also made. The apprentice question approached a solution as certain clauses of the agreement also included that category. The workers' struggle against the grade structure of skilled workers met with a certain degree of success although this was not complete. Disputes about the interpretation of this term were to be dealt with by a special trade council with representatives of both workers and employers and an impartial chairman. This council was also to investigate the apprentice question.[47]

Gains were also made in the problem of the piece wage paragraph but these looked a little more modest. Already during the first stage of the negotiations when the issues were being prepared by subcommittees, a large number of the workers' demands were deleted. The subcommittee on piece wage questions proposed the changes later to be found in the new agreement. In addition this committee noted that the workers required changes in the preamble to clause 2, that fixed schedules "should" instead of "may" be set up. The employers firmly opposed this demand and it does not appear to have been discussed in the conciliation offer.[48] The important gains in the piece wage paragraph were thus made at the beginning.

A concession had been made by the employers with regard to clause 2 where it had been laid down that piece wages could be altered independant of the notice period for fixed schedules where a change in method or machine was made that altered the production capacity of the workers. It was now admitted that routine reference to changed conditions did not justify a lowering of the piece wage but that it must be possible to show that the changes really caused alterations in earning capacity. This was thus a step in the direction of increased security or, to put it another way, a way of preventing misuse of the earlier provision.

The second paragraph of clause 4 was altered in two ways. "the worker concerned" was changed to "the worker". In accordance with paragraph 1 of the agreement the latter expression could be taken to mean "workers' organization", so that the change carried out, at least according to the Metal Workers Union, gave the trade union organizations the right of insight that earlier only the workers directly concerned had had. In a similar manner "time and motion studies" replaced the expression "result of time and motion studies". The more comprehensive expression increased the provision's area of application.

The following paragraph was added to minute 4, which had been inserted at the previous agreement:

> Should circumstances arise with regard to any plant, which suggest that the employer is offering unreasonably low piece wages—for the purpose of, as these are unacceptable to the workers, having the work carried out solely on hourly pay—the Metal Trades Employers Association is prepared to arrange for an investigation and, if there is reason for this, to work for a more correct setting of piece wages.

The idea behind the provision was to strengthen the workers' position in negotiations prior to a new piece wage. It would not be possible for the employers to frustrate the negotiations with an unreasonably low offer and thus get the work carried out on hourly pay, which even with the piece wage compensation was often lower than the piece wage pay. An arbitration of 2 September 1912 between the parties laid down that disputes about piece wages could not lead to industrial action but that in such cases the work should be carried out at hourly rates. The workers had long demanded that all piece wages should provide at least time and a half: unreasonably low piece wages were thus in breach of the agreement.

The iron and steel trades agreement of 1937 has been described as setting a standard for other agreements in the collective negotiations during that year. It was precisely on the point in question that the iron and steel trades agreement contained an innovation that irritated the Metal Trades Employers Association considerably. It was laid down in paragraph 2 clause 2 of the minutes that if the employer introduced new working methods or improved machines either party should be entitled to terminate the piece wages concerned within one week. From the time the piece wage ceased to apply the worker should be paid an hourly wage equivalent to the total average earnings achieved for the work in question during the last twelve months. The employers should make a proposal for a new piece wage and negotiations about this should be started immediately. If no agreement was reached within a month the employer was entitled to reduce the pay by 10%. If a further three months passed without agreement a reduction of a further 5% could be made. The parties were enjoined to cooperate willingly for a rapid and objectively correct setting of the piece wage.[49]

In the opinion of the Metal Trades Employers Association the provision was unsuitable and the association feared that the concession in this area made by the Iron and Steel Works Association would be utilized by the other side to obtain gains in other areas.[50] The new construction of minute 4 of the metal trades agreement should be seen as a variation on the same theme, where the union did not get so far.

Following the agreement there was satisfaction in the Metal Workers Union with the results. In spite of this irritation and differences of opinion about the interpretation of the altered piece wage paragraph were to arise in the following year.

5. The Metal Workers' Union Congress and Time and Motion Study Committee

With the agreement signed in January 1938 it became increasingly clear that time and motion studies had come to stay. They had been acknowledged as the concern of the workers' organizations, not solely as a tool for the employer in his relations with the individual worker. As the union now had the right in disputes to be informed of the grounds for time and motion studies the demands on the central organizations' theoretical consciousness of time and motion studies increased. Time and motion studies also became more usual and this increased the experience of both parties with them and at the same time created greater annoyance on both sides.

This development was clearly expressed at the ordinary Congress of the Metal Workers Union in 1938. No fewer than 8 motions were about how the union should deal with the aftereffects stemming from time and motion studies.[51] Some proceeded from practical experience on the factory floor and others from the possibilities and responsibilities brought about by the new agreement. Most of the motions resulted in demands that the executive board should accept responsibility for increased information to the committees of the branches and local unions as well as to individual members. This could be done by courses, study circles, travelling instructors or in some other way comparable with the union's other study activities. But many motions avoided any definite demands and intended instead to set up an investigation in the union to discover alternative ways of action.

In its statement on the motions the executive board fastened on the last idea. Initially the union's principal attitude was again confirmed:

> Time and motion studies are often an instrument for the achievement of rational production. Increased efficiency in production is a condition for the worker to obtain greater pay and thus an increased standard of living for himself. To oppose rationalization and the associated time and motion studies thus means that the worker opposes the conditions for increased pay.[52]

The board continued by defining the role of trade union organizations:

> It lies with the trade union movement to protest and prevent the worker, by means of rationalization and time and motion studies, being forced to carry out unnatural performances by which he can suffer irreparable injury, and also by action about pay and working hours to try to ensure the workers the greatest possible share in the profits stemming from rationalization.[53]

The executive board's recommendation to Congress with reference to the motions was to instruct the board to appoint a committee of five people for the investigation of which measures ought to be taken in order to provide the union's members with a greater insight into time and motion study questions and also of the contractual rights that exist in the area; that study circles on the subject of time and motion studies receive grants from union funds in accordance with the same norms that apply to other approved subjects; . . .[54]

The debate, in which about twenty speakers joined, revealed the whole spectrum of attitudes about time and motion studies. There were those who were pleased to observe the executive board's positive attitude as the workers could not be permitted to get left behind in relation to the employers when it was a case of utilizing time and motion study techniques in the workers' interests.[55] There were those who, like the board, regarded time and motion study as sometimes a necessary evil but in any case an actual fact. There was no going back; if the trade union had accepted the piece wage system it was not possible to protest because the piece wage was fixed on more scientific grounds.[56] There were those who took their stand on time and motion studies according to the purpose for which they were utilized. Frithiof Thapper from Finspång (then a young man but later to end his career as Speaker of the Swedish *Riksdag*) thought that time and motion studies as a means of increasing production in order to increase the workers' pay was a good thing, whilst they could not be accepted as a means of increasing profits; the allocation of financial results moved to the fore.[57] Finally there were those who, without making any request, simply took the opportunity of expressing their dislike of rationalization and time and motion studies in all shapes and for all purposes.[58]

Congress approved the executive board's proposal with one addition, which heightened the demand to obtain rapid and concrete results from the investigatory work:

that the investigation committee is instructed to present at the next policy conference proposals for an extended right for the workers in time and motion studies that the investigation may lead to.[59]

Soon after Congress the executive board commissioned Erik Collin to work out in detail how an investigation committee should be made up and what tasks it should be given. His proposal was placed before the board's meeting at the end of September when the arrangement and the names proposed were approved. The committee was to consist of, in addition to Collin as chairman, three representatives of the metal trades, and one for the iron and steel trades. Their task was to be:

to work out proposals whose realization is devoted to providing the union's members with a broader insight into time and motion study questions and about their contractual rights that are to be found in the area.[60]

The committee members had a clear time limit for their work. The proposals were to be presented to the very next policy conference in September 1939. The committee was able to present its report within the time laid down.[61]

The committee began by describing its work and its contacts. It had met for three periods when it had carried out study visits to L. M. Ericsson, Stockholm, SKF, Gothenburg, and Borås Wäfveri, a textile mill. In November 1938 the committee had been invited to discussions by the committe on human engineering of the Royal Academy of Engineering and Sciences.

At these meetings the committee obtained certain valuable information that has been advantageous in the decisions of the committee on various matters,

said the time and motion committee of the Metal Workers Union with some degree of satisfaction.[62]

The committee then went on to present proposals on how information activities could be arranged. Lectures were one of the best channels for information at the instruction courses or branch meetings arranged centrally or by districts by the union. Special courses of instruction about time and motion studies only were, however, not considered necessary as yet. For the lectures the committee recommended that the union should produce short films and study material for the metal trades industry and the iron and steel industries. With this end in view contacts had already been made with some companies. The making of a film had also been discussed but the idea had been dropped. It would certainly cost more than it was worth.

The study circles should be encouraged, as well as Brevskolan (The Correspondance School, one of the educational undertakings of the joint Labour movement) where a series of study letters had already been prepared. Contacts had been made with the central office of the Swedish Workers Educational Association (Arbetarnas Bildningsförbund). During these the need for literature and a study plan had been realised. However, the committee had had the chance of looking at Tarras Sällfors' "Arbetsstudier inom industrien" (Time and motion studies in industry) in proof and suggested that the committee should be instructed to produce a study plan on the basis of this book. The union should also help in a cheap distribution of Sällfors' book. Sällfors was the leading industrial consultant of the period so thus the same educational material was used by employers and workers alike.

The committee had also had the task of preparing the draft of texts that could be inserted in the agreement for the purpose of giving the workers an extended right in time and motion studies at workplaces. Here it had proceeded from the basis of the current agreement and presented proposals for such additions.

The committee's proposals were wholly accepted by the executive board. The texts formulated for the agreement were thus submitted to the 1939 policy conference without any changes. The other proposals were passed to Erik Collin to follow up.[63]

Later in the autumn of the same year he was able to propose to the executive board to buy 500 copies of "Arbetsstudier inom industrin", at a price subject to rebate, to let the consulting engineer Harald Ager, suggested by Sällfors, work out the technical sections of a study plan based on Sällfors' book, and himself to construct the part dealing with agreement questions. The intention was that members of the union should be offered Sällfors' book at a price still further reduced and that the study plan would be distributed without cost to the members of study circles. The main lines of these proposals were also approved by the executive board.[64] The decision to purchase Sällfors' book was repeated a number of times during the next few years; altogether the union bought at least 2,500 copies of the first three editions.[65]

Taken as a whole the work of the time and motion study committee meant that the union and its members were considerably better equipped than earlier for a reality where time and motion studies constituted an inevitable part. The committee's proposal about the wording of the agreement was to decide the union's demands in many collective negotiations. The union placed at the disposal of its members, the branches and the districts, various kinds of information material and funds for the support of study circles. The Metal Workers Union thus acquired a theroretical foundation of the same origin from which the employers also proceeded. The time and motion study committee's satisfactory collaboration with the committee of the Academy of Engineering and Sciences must be remembered as also the utilization of Sällfors' and Ager's services. To a certain degree this collaboration was caused by the fact that no other authorities were available but it is also a sign that the trades union movement accepted the idea of the rationalization movement; that a comprehensive, common, technical foundation existed for time and motion study methods and that this was not a matter for argument between the parties. The use of Sällfors' book was not free from internal criticism among the leadership of the Metal Workers Association. Collin's answer to this was that it was not necessary to read the book uncritically.[66]

6. Agreement problems during the war years

For the two sides of the metal trades industry the collective negotiations of 1937 meant the commencement of a period of firmer direction from the main organizations. In 1939 the control exerted by both the Swedish Employers Confederation and the Trade Union Confederation became even firmer because the outbreak World War II stimulated national solidarity although Sweden remained outside. The guiding lines to be followed in the collective negotiations were drawn up by the Trade Union Confederation at an information conference at the beginning of September.[67] Once again preliminary negotiations were recommended and modest demands, mainly guarantees about the wage level during the period of the agreement.[68]

The iron and steel workers' policy conference on the agreement was held on 18 September and the metal workers' policy conference on 25–26 of the same month. For both areas of agreement two alternative proposals for agreement were approved; one to be used at the preliminary negotiations, the other to be presented if notice was given. The first proposal involved important demands for the increase of hourly pay and piece wages by 5 %. Changes in the piece wage paragraph were not claimed. On the basis of this proposal negotiations were carried on and agreement was reached on 2 January 1940. During the intervening period was born the first framework agreement between the top organizations by virtue of which wages on the Swedish labour market were in principle linked to the cost-of-living index.[69]

The second proposal, which was approved by the policy conference at the end of September, is more interesting in the present context. It contained an initial more ambitious attempt to deal with the time and motion study problem. The text of the agreement in this case—paragraph 4 clause 4—resulted from the work carried out in the time and motion study committee set up by the decision of the union's Congress in 1938.

The proposal left clauses 1–3 unchanged as well as clauses 5–8, the latter having changed numbering as the provisions concerning bonus piece wages were again detached and placed in a special clause, No. 5. The time and motion study provisions were found assembled in clause 4:

> In those cases where the piece wage prices are calculated in accordance with time and motion studies, agreement shall be made on the grounds of calculation, which likewise include the supplementary percentage for lost time. The workers have the right to receive copies of the time and motion study report as well as having a voice on the length of time or the number of pieces which the time and motion study shall cover. With regard to the information contained in the time and motion study report, agreement will be made between the local parties. The workers, by which is meant also the representatives of the workers' organization, have the right during time and motion studies to investigate that they are being carried out in the agreed manner. When a piece wage agreement is made on the basis of diagrams or tables prepared, the workers have the right to receive copies of these together with the basic material for them. In such cases agreement shall be made on the grounds of calculation and the additional percentage for lost time. Disputes that may arise in the setting of piece wages in accordance with the provisions in this section shall otherwise be dealt with by reference to the provisions concerning negotiation procedure in the agreement.[70]

Even a glance at the scope and content of the section shows that the workers were prepared to set about regulating the technicalities surrounding time and motion studies in quite a different way than earlier. This preparedness is an expression both of a knowledge and a desire and likewise implies an acceptance of time and motion studies right down to the technical details.

The exclusive right of the employers over time and motion studies was to be undermined by this proposal in certain important points. The employer was still presumed to have the unrestricted right of decision on whether time and motion studies should be carried out. But when such a decision had once been made the workers (and their organization) were to have a voice in the scope and carrying out of the time and motion study, on what the report was to contain and on how the data gathered were to be handled, that is to say, on the basis of calculation (cash factors) and the size of the lost time bonus. Local agreements were to be made by the local parties about these as well as on the writing of the report. Other kinds of material used as a basis for the calculations of time and motion study results were not to be withheld from the workers. Thus local agreements were to be made about more general conditions whilst in the individual case the workers were to be guaranteed influence on how agreement was reached. The workers—and their organization—were to have the right to verify that time and motion studies were carried out in the correct manner.

No great discussion arose on this proposal at the policy conference. The speeches made were less intended to criticize than to stress the extent of the problems that had also led to the executive board's proposal. A reading of the minutes of the meeting produces an impression that the matter was discussed rather in passing. Most of those present were presumably awaiting preliminary negotiations and so for the alternative propsal that implied an unchanged piece wage paragraph. At later policy conferences on the agreement there were in fact representatives who had not noticed that this proposal of the time and motion study committee had been submitted.[71]

The first framwork agreement of December 1939 was followed during the war years by a further four, concluded in December 1940, January 1942, December of the same year, and January 1944.[72] These framework agreements expressed the increased central direction from the top organizations under the pressure of Sweden's exposed position. Mainly they were to function as guarantees that the expected inflation should not have too severe or excessively rugged consequences for the wage level of the workers.[73] Room was thus left for the individual unions to negotiate on alterations of the current agreement in other areas. The ability to advance other issues of fact was, however, limited: the conflict weapon could hardly be considered available following the conclusion of the framework agreements so long as the world situation still looked so dark. The agreement arrived at in January 1940 was therefore prolonged three times so that it remained valid up to the end of 1943 with certain alterations mainly caused by the framework agreements.

Policy conferences on the agreement were held in the usual way, however, with the workers of the metal trades industry in September of 1940, 1941 and 1942. On all three occasions the conferences decided that the changes in paragraph 4 clause 4, which had appeared in the second alternative proposal in 1939, should be put forward in the negotiations should the agreement be terminated. Nevertheless the dissatisfaction increased from conference to conference. In 1942 it was proposed that changes in the piece wage paragraph should be introduced at the negotiations with the Metal Trades Employers Association about the prolongation of the agreement.

But his more stringent attitude was rejected by the conference with a vote of 49 to 37.[74]

So it can be seen that there was no change in the provisions of the agreement on time and motion studies during the first four years of the war. But there was increasing annoyance on the workers' side which was undoubtedly also connected with a mobilization in the employers' camp. The Metal Trades Employers Association took a much greater interest in the question than it had earlier.

7. The Metal Trades Employers Association activated

In the earlier text has been shown how the Metal Trades Employers Association started to deal with the question of time and motion studies by appointing personnel to study the development specially and by formulating a line for the association to take when it was necessary to decide how piece wage calculations should be carried out. There was an increased preparedness on fronts that were becoming active in the trade union struggle.

A the same time the association was reluctant to engage itself too deeply and in a practical way. At a conference of the southern branch of the association the need of employing a consultant time and motion study engineer had been expressed. The small companies were unable to pay the fees charged by consulting firms like the Industrial Office Ltd (cf. p. 53). The problem was now whether this branch could employ someone. This illustrates how the board of the Metal Trades Employers Association regarded the part played by the association. The help that small companies undoubtedly needed could not be given suitably by the association, declared Georg Styrman. The association was the final authority in disputes about piece wage questions and the association's time and study man could not be tied down beforehand. On the other hand perhaps somebody could be appointed through the Swedish Association of Engine Makers (Sveriges Maskinindustriförening) or through a number of companies in the local branch together.[75] And in this way the matter was finally arranged.[76]

At the beginning of the 1940s the level of activities again increased, not least this time because of trouble in its own ranks. At the end of January 1941 the board of the Swedish Metal Trades Association decided to arrange a gathering at a central conference of time and motion study men who were active in member comapnies.[77] This took place on 19–20 February and a varied mixture of information on factual questions and policy were laid before it. During the discussions it was stressed from many quarters how important it was that conferences of this kind were held in order to coordinate the treatment of piece wage problems. The managing director was therefore instructed by the board to arrange local conferences.[78] These were held during April and May in five districts.[79] The overall total attending was approximately 500.[80] According to the report to the board the conferences had been successful, not least because a great deal of valuable material had been gathered that could be utilized as evidence in future actions on piece wage questions at the Labour Court. At the conferences it had been hoped that similar meetings should take place regularly.[81]

Further, the question had been raised as to whether the Metal Trades Employers Association should not create a centre for the purpose of coordinating the results of time and motion studies. The point was put to the association's apprentice committee, which had a meeting at the end of May. If the association did not create a committee to look after the interests of the employers in the metal trades in connection with time and motion studies, it said in a memorandum submitted, there would be a great risk that some other institution would take up specific employers' questions. There would not be a lack of tasks for it. The committee could start by dealing with the "false doctrines" being spread by the Institute for the Training of Foremen (Arbetsledareinstitutet)—see below. Further, the committee could arrange new time and motion study conferences, make contacts with time and motion study men in the field, and follow-up the time and motion studies of the members' plants.[82]

The apprentice committee, which for long had functioned as a technical committee within the association, agreed completely with the proposal that a committee should be created. "As a counterbalance to the existing time and motion study committees", as Erik August Forsberg expressed the purpose. Hellman from ASEA pointed out that the Institute for the Training of Foremen and its courses for time and motion study engineers had taken up questions of pay in such a manner that ASEA had ceased to send its personnel to the institute.[83]

The board decided at once, per capsulam, to set up a committee composed of Erik August Forsberg as chairman, Gösta Ingelman as secretary and five members.[84]

The committee held its inaugural meeting at the beginning of the autumn in 1941. In a memorandum Forsberg stated the motives behind its formation. For many years there had been a desire that the association should be able to satisfy its specific interests in questions relating to time and motion studies. One way was to work together with the institutions already established in the field. The demand for an initiative from the association's side had been given currency for a number of reasons. The Swedish Metal Workers Union had issued a study guide to Sällfors' book on time and motion studies. Forsberg considered this contained gravely misleading information. The Institute for the Training of Foremen spread wrong ideas at its courses about the effects of rationalization and the workers' right to share in the profits of rationalization. Furthermore there were fears that a similar unsuitable activity would be developed at the "Comrades Club" that had been formed at this institute by a number of time and motion study engineers from different trades for the purpose of discussing matters of common interest. Finally, it was a positive desire from the members that possibilities should be created to gather and exchange experiences about time and motion studies.[85]

Forsberg's account only suggests the conceivable reasons for the association taking up the question of time and motion studies precisely at the end of January 1941. The metal trades agreement, with regard to time and motion studies, had not been changed since 1938. The Institute for the Training of Foremen had been active for a long time and the need for members to exchange experiences could not have been a sudden idea either. It is quite certain therefore that it was partly the study guide of the Metal Workers Union, published in August 1940, and partly the establishment of the "Comrades Club" that triggered the action of the Metal Trades Employers Association.

Already at the inaugural meeting a number of future fields of activity for the time and motion study committee were presented. The question of time and motion study conferences and the training of time and motion study men was to be the object of the committee's continual care. Cooperation with other institutions was also to be considered from time to time. "Cooperation" in this context seems to be a euphemism: what it was all about was to get the others to dance to the tune of Metal Trades Employers Association or to leave the floor. The committee was also to devote its energies to questions about the agreements as they arose in negotiations. It was also necessary to be on the watch for and to gather evidence that could be used in possible future action in the Labour Court about the possibilities of earnings and other disputes.

The idea that the association should create an information centre for time and motion studies had already been discussed at the conference on this subject in the spring of 1941. Among a number of others Hjalmar Nilsson (Bolinder-Munktell) suggested that the association, in line with the German pattern from Refa, should collect standard times for suboperations that occurred frequently and put these at the disposal of the members. But when this was presented in committee the majority adopted a sceptical attitude: standard times took no account of the fact that every workshop had special conditions. In addition standard times had a tendency to create fixed schedules. The committee decided to urge the proposers of the motion to state exactly how the proposal should be carried out in practice.[86] Nilsson later complied with the request but the committee did not depart from its wait-and-see attitude.[87]

The time and motion study conferences were held annually following the good experience obtained in 1941. There was none in 1944 for special reasons. Usually the conferences were held by branches in five districts. A standard program was used, where some case studies were presented, in the context of Styrman's and Ingleman's lectures about agreement questions and methods of calculating piece wages. A general discussion followed. The main weight of the program undoubtedly lay on agreement and calculation methods, not on time and motion study procedures as a technical craft. In this way the conferences served as a mustering of members of the association, with possibilities for the centrally employed officials to take the pulse of those in the field and to influence them in the required direction. The conferences also functioned in this way as a reference group for the time and motion study committee in its function as a drafting committee on questions concerning the agreement. Thus the conferences played a part in the policymaking of the Metal Trades Employers Assn. and in suitable cases they will be considered in this context.[88]

That the association should initiate courses for time and motion study engineers was explained by the existing shortage and not initially because of dissatisfaction with the training carried on under the management of the Institute for the Training of Foremen.[89] In fact the association linked up with this institute which from the spring of 1942 gave special time and motion study courses twice annually for the metal trades industry, partly paid for by the association. Instruction was based on Sällfors' book and on Ingelman's articles in the journal Verkstäderna (The Workshops) which were published subsequently from June to October 1941. Each course lasted three to four weeks and was largely devoted to time and motion study methods, being led by Sven Engelgren and Harald Ager, whilst piece wage and agreement problems received a smaller proportion of the time. The lecturers for the last two subjects were the Metal Trades Employers Association's own officials Styrman and Ingelman.[90] The first and most important courses were aimed at time and motion study engineers. Later there was also a certain amount of training for time and motion study men, that is to say the engineers' assistants who were normally trained by them.[91] Towards the end of the period there were also shorter continuation courses.[92]

Collaboration with the Institute for the Training of Foremen had started under inauspicious conditions so it is not particularly surprising that this was broken off in the autumn of 1944. A certain amount of relief marked the discussions of the committee about the decision that the association should follow its own path and, without any collaboration from the training institute, arrange the course for the autumn.[93] This was an indication as good as any that at that stage the association intended to carry on a program on the time and motion study problem fully and in numerous ways. It had now become a matter for the organization and not the individual company.

This initiative from the association came, however, in a field where there were already many interested parties active. The unavoidable result was collisions and demarcation problems.

Suspicion of other organizations and institutions who entered what was considered by the Metal Trades Employers Association to be its own area has already been shown quite clearly by the earlier account of the genesis of the time and motion study committee. Through this committee the association, in relation to other institutions, was to establish the views and interests of the metal trades industry.[94] Forsberg's idea was, however, that they should take it easy from the beginning without taking the initiative.[95] This line was followed for a year except in a case where collaboration was arranged with the Institute for the Training of Foremen on time and motion study training. But in the autumn of 1942 the time was ripe for an offensive against the Academy of Engineering and Sciences and its time and motion study committee and against the earlier-mentioned "Comrades Club" at the foremen training institute.

During 1941 the "Comrades Club" had been reconstructed to become "The Association of Sweden's Job Study Managers".[96] At the spring course for the job study engineers the principal assistant at the training institute, Sven Engelberg, had urged the participants to join this association. Faced by the threat that an association dominated by Engelberg and the institute should expand and impinge upon the employers' interests, the management of the Metal Trades Employers Association went into action. To nip matters in the bud a memorandum was sent out to the members of the time and motion study committee. It was not rational, it was asserted, to have an association for job study engineers that covered many branches. Job studies always touched upon specific questions for the branches. As these questions were often closely related to questions about agreements the development was still more serious:

> It cannot be permitted that such association shall on their own deal with questions associated with the agreement. With regard to, for example, the agreement for the metal trades industry, it was easy to meet problems of interpretation that were extremely difficult to decide, not to say subtle, that the functionaries concerned in the Federation of Swedish Industries' Institute for the Training of Foremen could not settle for the job study men of the metal trades industry . . .[97]

An association such as that urged by Engelgren for Sweden's job study managers was unwelcome. On the other hand discussions at the job study courses had undoubtedly reflected the need for some kind of association. The leadership of the Swedish Metal Trades Employers Assn. therefore recommended that an association should be formed solely for the job study engineers of the metal trades industry. "Some suitable people from the metal trades industry could be requested to head the association in question", it continued.[98] Control over the exchange of ideas would be ensured for the Metal Trades Employers Association and the Association of Sweden's Job Study Managers would be driven out of the area.

The members of the time and motion study committee thus had to form an opinion of the memorandum they had received. Three members agreed with the recommendation to form a separate association but two were prepared to accept the Association for Sweden's Job Study Managers. In his positive approval one of the two requested that the chairman of the existing association should be allowed to give his opinion on whether the risks described really existed and continued:

If the existing association only includes job study managers and in addition only a limited number, it must be presumed that the members are of such a high class and so responsible that the risks are not especially great. For they are in their daily work responsible for the setting of wages and must certainly on this account be considered so expert that they should not require the assistance of functionaries of the Metal Trades Employers Association. From the impression I have obtained the existing association should not have anything to do directly with the Institute for the Training of Foremen.[99]

At the meeting of the time and motion study committee in October 1942 the question was thus whether the Metal Trades Employers Association should persist in forming a separate association of job study engineers within its own area of activities, or whether it should give up the idea of being the guardian and let the matter drop. The committee reached no decision but Erik August Forsberg, who had the last word, considered that the association ought to keep in touch with the development without taking any direct initiatives.[100] At a later date the committee discovered there was reason to go in and influence the planning of the activities of the Association of Sweden's Job Study Managers.

It was also necessary for the committee to deal with a memorandum about the duplication of work in the field of vocational training and job studies. In the spring of 1940 the Federation of Swedish Industries had taken the initiative for the emergency training of semiskilled workers. This was something that applied mainly to the requirements of the metal trades industry, and the Metal Trades Employers Association had already started such training some months before in collaboration with the National Labour Market Commission. The Academy of Engineering and Sciences was about to take up the recruiting problems of industry in spite of the fact that both the Metal Trades Employers Association's apprentice committee as well as the Labour Market Committee were already dealing with the question. It was certainly important, said Georg Styrman, that duplication of work was avoided:

The employers had commissioned the Metal Trades Employers Association to deal with certain tasks so that it would then be unsuitable that other association started to deal with the same questions without collaborating with the Metal Trades Employers Association.[101]

During the speech about rationalization and avoiding the duplication of work the irritation could not be concealed. There was a strong mistrust in a definite direction: the Institute for the Training of Foremen and especially its director, Gösta Ekelöf, were the common factors in the initiative criticized. The committee decided to deal with the problem by contacting the Academy of Engineering and Sciences. The sharp-tone of the proposed communication was considered by Erik August Forsberg to be too strongly worded. The closing remarks, which were deleted, produced a stark impression of the demand for guardianship that the leaders of the association wanted to exert:

In this connection we should like to point out that in your Communication No. 119, certain statements were published that might possibly cause inconvenient consequences ... as piece wage questions are extremely complicated we consider that one should be extremely careful in the publication of statements that might be misunderstood or generalized. We should therefore feel greatly obliged if from now on and prior to publication of such statements you would let us have sight of them.[102]

The Metal Trades Employers Association's proposal to the Academy led to a discussion on 16 November 1942 and finally to a conference under the auspices of the Employers Confederation in February 1943. The result from the Metal Trades Employers Association's point of view proved acceptable. Styrman was able to note statements from the representatives of the committee of the Academy of Engineering and Sciences that in fact meant submission to the demands for correction made by the Metal Trades Employers Association.[103]

In the autumn of 1943 the question of the relationship with the Association for Sweden's Job Study Managers came before the committee once again. Patrick Rydbeck reported that there had been a great deal of discussion within the association about the differences revealed at the conference of the Employers Confederation in February. The job study managers association, Rydbeck said, was now preparing a conference on collaboration between the Employers Confederation and the Trade Union Confederation about time and motion study problems. This initiative came as a surprise to the Metal Trades Employers Association said Erik August Forsberg, and supported Rydbeck's proposal that Gösta Ingelman should apply for membership of the Association of Sweden's Job Study Managers in order to be able to follow future developments.[104]

Increased vigilance soon turned into action. At the next committee meeting Styrman reported that in a conversation with a functionary he had straightened out the Association for Sweden's Job Study Managers with satisfactory results:

... In the discussion were advanced the misgivings of the Metal Trades Employers Association about the arranging of such a conference by private associations in the present situation. Were such a conference to be arranged the initiative should come from more authoritative quarters. In view of the negotiations and the plants' difficulties in releasing personnel in the present circumstances for such conferences the time selected was not well-chosen. The planning committee of the Association for Sweden's Job Study Managers had later reported that the conference was to be postponed for an indefinite time.[105]

Once again the Metal Trades Employers Association had repulsed an encroachment into it own area that threatened the association's control over the development.

The setting up of the employers association's job study committee was the result of increased activity in other quarters of trade. The association was forced to move still further away from the passive attitude to job study questions it had adopted as a principle. The birth of the job study committee also indicated an increased ambition and readiness on the part of the association. This showed itself in the future collective negotiation activities where the committee was able to act as a technical committee within the association with greater certainty, and also in relation to other organizations and institutions. The committee's task of maintaining the interests and views of the Metal Trades Employers Association—the employers in the metal working industry—led it to make demands for adjustment from other groups in the trade. Especially those who tried to pursue a line between the main parties of the labour market discovered how the space for action had been drastically reduced. Another expression of the association's increased ambitions was the commitment to training and active information measures through which the attitudes of the Metal Trades Employers Association could be projected and lead to the creation of norms in wider circles.

8. Summary

In the above it has been possible to show a change of policy in both the Metal Workers Union and the Metal Trades Employers Association with regard to rationalization and, especially, job study questions.

The old tactics of the Metal Workers Union in the job study question, bearing Ekman's signature, namely to try to obliterate the differences between piece wage categories, was carried further under the chairmanship of Gunnar Andersson but was abandoned until the late autumn of 1937. It was replaced by a line where the importance of the fixed piece wage was very strongly stressed and where the temporary piece wage was dealt with as an exception. There was a constant aim: to secure wages by including all piece wages without question under minute 1.

The causes of the change in the union's demand are difficult to explain unambiguously. Apparently the decision of the Labour Court was of importance here as it clearly confirmed the existence of fixed and temporary piece wages. This division emphasized the union's new line: the influence of the *local* trade union organizations in a position where there had been a lack of success in upholding the wide validity of the *central* provisions. The change of chairman in the union may also have been of some importance. From the end of the 1920s Oscar Westerlund belonged to the acid critic's of Ekman's policy and already then defended fixed piece wage.

Per Erik Back has characterized the period up to the middle of the 1930s in the history of the union as "defensive". In view of the attitude to rationalization and job studies perhaps the word passive should also be added. The union took no definite steps of any kind to come to grips with time and motion study problems. The Labour Court had certainly implicitly made piece wages into a question for the union in 1934, but the agreement of 1936 still regarded job studies as a matter between the employers and the individual worker. Not until the 1938 Congress and the time and motion study committee then appointed, did the union become more ambitiously involved in the job study question and obtain a considered program. This was marked by the same ideas that also guided the Trade Union Confederation's policy: responsibility for rationalization and the demands for joint determination thus caused. For itself and its members the union started to demand that they should be able to utilize job studies for their own purposes.

The Metal Trades Employers Association was also compelled by the development to give up the passive stand on rationalization questions it had adopted earlier. To coordinate its own policy and understand the reality out in the plants and in order to maintain the association's interests, not least in relation to other organizations rather close to the industry, it set up a permanent organ. A further step towards greater responsibility for job study questions was taken by the association when it took under its wing the training of job study men for the metal trades industry in 1944. The build up and the preparations of both negotiating parties were subjected to a delaying effect as World War II caused increased centralization and the "freezing" of negotiation in the labour market. Not until the war situation turned during the winter months of 1942/43 were the latent energy and industrial development and the increased preparedness of the organizations to find their release in confrontation on questions about piece wages and job studies.

CHAPTER 5

Conflict and concord

1. Collective negotiations 1943

During 1943 the workers' dissatisfaction increased, not only with the pay freeze, which could not prevent a decrease in real wages, but also with a number of other things including the job study question.

At the policy conference on the agreement in September the executive board of the Metal Workers Union proposed that the date for giving notice to terminate the agreement should be deferred in order to create time for preliminary negotiations. At these the expected framework agreement should be presumed and a number of other demands put forward.[1] In the piece wage question the proposal for a new construction of clause 4 that had been in abeyance since 1939 should really be urged with additions to strengthen the workers' side still further, such as the right to take the initiative for job studies. The clause should begin in the following way (additions in italics):

in those cases where the piece wages are calculated *and proposed* following time and motion studies, agreements shall be made *before the work begins* ... between the local parties. *If the workers at a company, which has introduced time and motion studies, require that a certain job shall be subjected to time and motion studies, this shall be done.* The workers, by which ... etc.[2]

The union's chairman, Oscar Westerlund, was slightly optimistic and thought that increased protection in the agreement against the pressure of the job study men should not entail any direct obvious costs for the company so that the requests put forward should be possible of achievement.

At the continued discussions at the policy conference it was emphasized that the job study question had become increasingly urgent. Many speakers demanded that notice of termination should be given in order to increase the weight of the workers' arguments. One of the most expressive of these demands came from a deputy of Södertälje:

The inquisitorial methods of the middle ages have found their renaissance in the factories. Piece wages are squeezed unreasonably ...

To enter into preliminary negotiations did not decrease the weight of the demand for changes in the piece wage paragraph, promised the representative, Erik Collin. He confirmed that the job study question should "be settled whatever happened" at the impending negotiation.[3] The same determination was exhibited by Lennart Eckerström, a member of the executive board:

In the proposal made by the board there is one special question that is important and that is the question of the piece wage paragraph. I do not regard this question as one between us and the Metal Trades Employers Association, but a question that concerns the whole of society. Trade cannot function as it should because of certain disturbances, which have their source in the formulation that the piece wage paragraph now has.[4]

The executive board thus demonstrated a greater fighting spirit within the framework of the proposed preliminary negotiations. The piece wage paragraph particularly was to be brought into the line of battle. With this arrangement the board's proposal became acceptable to the conference with a good two-thirds majority.[5]

The time and motion study committee of the Metal Trades Employers Association made its decision about the union's demands at the end of October.[6] The general reaction was very negative. George Styrman stated:

An approval of the present proposed alteration by the workers' side should mean that all factors that decide the piece wage became fixed. In actual fact this meant that the workers alone could decide the size of the piece wage. Presumably the result of this would be that we were compelled to abolish the whole job study system in the setting of piece wages.

Sentence for sentence, demand for demand, the committee then went through the Metal Workers Union's draft for agreement. The demand for central agreement on the grounds of calculation was rejected. If companies were prepared voluntarily to make agreements about the cash factors and lost time bonus, they would not be prevented from doing so. On this point the committee was compelled to fall back on the questionnaire investigation that the association had carried out among its members which showed that a surprisingly large number, at least internally calculated with fixed cash factors (cf. pp. 39 et seq.). At the earlier discussion on principles the members had unanimously rejected a freezing of the cash factors. Only Georg Styrman thought fixed cash factors could be an advantage "from the operational point of view"!

The other demands for changes were flatly rejected with the exception of the right for the workers' representatives to check whether a job study was carried out correctly. Patrik Rydbeck had had good experience of this from SKF (the Swedish Ball Bearing Company), whilst Sten Tiberg from L. M. Ericsson was of a diametrically opposite opinion. Georg Styrman considered that "at least initially attempts should be made to avoid consent on this point also", a phrasing that from Styrman sounds more like an invitation.

Apparently the negotiations on this point were very tough. The conciliator, Olle Ekblom, who was soon brought in, attempted a solution by avoiding changes in the text of the agreement and instead tried to get the parties to agree that the employers should issue a statement on the job study question. From this it was to appear that the association did not oppose that the worker concerned obtained influence on the extent of the individual job study, it did not object to the local union branches seeing the results of the job studies and that the workers should have the right to propose job studies; also the association did not see any objection in principle to agreements about the cash factor being made. But in the last question the employers could not be obliged to make agreements.[7]

The proposal, presented on 30 November, was turned down flat by both sides, but the following day Ekblom returned with a new draft with the texts of the paragraphs in the agreement unchanged but with five minutes attached. Four of them dealt with the same items that were to be covered by the employers' statement in the earlier proposal. To add these as minutes was to go some way towards meeting the workers' demands. In return was suggested a fifth minute that would lay down the responsibility of the workers to assist loyally in job studies.[8]

The workers' side, which for reasons easily seen was likely to have a basic mistrust of minutes, did not accept this construction completely either. The following day the conciliator presented still another proposal where the provisions on the local union's rights were included in the text of the agreement. But even this did not bear any fruit. The workers' demand was confirmed in the five-point proposal handed to Ekblom on the same day. These maintained the workers' unconditional right to have job studies carried out and the right of access to all material included in such studies. This right fell automatically "to the parties stated in the negotiation procedure", that is to the trade union organizations also. Agreement should be made on the basis of calculation (the cash factor) and the percentage bonus for lost time.

At this stage of negotiations Olle Ekblom made a further concession to the workers' side. He presented a proposal which included a passage about the top organisation's right to complete information when a dispute arose or in control studies.[9] At that time, on 9 December, the negotiations were aborted. The reason for the failure was considered by the workers' negotiating team to be, among other things, that a solution of the piece wage problem had not been achieved.[10]

The Metal Workers Union reconvened the policy conference.[11] In his opening speech Oscar Westerlund spent a great deal of time presenting and explaining the position taken up by the workers' side on the piece wage question. As chairman he was able to say that some success had been achieved as the Metal Trades Employers Association's opposition to a closer regulation of the piece wage question by agreement had been broken. But on a number of points he was not satisfied with the conciliation proposals presented.

The organization, the Metal Workers Union, must, he said, be granted the right of access to job study material even when a dispute was not involved. The local branches and workshop clubs should have the right through their representative to follow the control studies carried out to resolve a dispute, and the expression "as soon as there is time" in this context provided the opportunity for an unacceptable arbitrary attitude. Workers and employers should further negotiate on the extent and execution of the individual work studies before they were started. The workers' right to influence the performance of such studies should be made automatic.

With regard to the calculation of piece wages many terms were used in the debate: "earnings factor", "cash factor", and "basis of calculation". The workers should try to get the first of these terms into the agreement. It could be so interpreted that the most recent earnings achieved by the workers, that is after the building up of a piece wage, should form the basis for new piece wages. The regulation of the lost bonus should be discussed. Finally the proposed demand in the minutes for the workers' loyal cooperation in job studies should be made applicable to both parties.

Westerlund's account resulted in the proposal that in spite of everything negotiations should be carried further and the negotiators should seek for improvements on the points mentioned.[12] The policy conference followed the chairman's recommendation because acceptable compromises were in the offing.[13]

At continued negotiations before the conciliator the Metal Workers Union obtained some sympathy for still more of its demands in the piece wage question. The local unions in the conciliation proposal of 17 December received the right to follow control studies through a representative. "As soon as there is time" was altered to "without unnecessary delay" and the minute about the cash factors received an addition, which recommended the parties to "pay special attention to lost time". At last the workers' tough stand had compelled the conciliator to meet their demands step by step. However, the proposal of 17 December was much too favourable. The employers leaned hard in the other direction and reversed the swing of the pendulum. In the conciliation proposal of the following day the paragraph about lost time had disappeared once again and cash factors were called "grounds for calculation". In return for this in the new draft "the worker concerned" was not included but "the parties concerned" who were to cooperate loyally for a correct result of the job study work.[14]

On the basis of the conciliation offer of 18 December the Metal Workers Union was again summoned to the reconvened policy conference. Two points especially where the aims had not been achieved were urged by Westerlund and Collin. The right to receive complete reports on job studies had not been guaranteed and the right to be present at job studies had been granted to the representative of the local union but not to the central organization unless the dispute had been carried to central negotiations.

The proposal for the agreement received a mixed reception at this conference. Many speakers considered the paragraph was worded to vaguely. But most of those present supported Westerlund who declared he could not "consider the results of the negotiations to be bad". With 69 votes to 36 the conference decided to approve the draft agreement, which was signed by the parties on 30 December 1943.

2. The Labour Court 1944

In the negotiations of 1943 there were real discussions between the parties for the first time about the techniques of job studies.

One of the points about which different opinions cropped up concerned whether the grounds for calculation in the fixing of piece wages should be the subject of regulation in the central agreement at all. A compromise solution had been arrived at in the agreement's recommendation that really meant nothing at all. This provides a good reason to spend a little time on the fundamental problems and the antagonisms they caused, which were undoubtedly responsible for a great deal of the irritation about job studies. A memorandum originating from the Metal Trades Employers Association's internal discussion on policy in 1942 provides a good starting point. The memorandum went straight to the point:

The question concerns:

1 Should a fixed cash factor be used in a plant for a group of workers or
2 should the job study man in consultation with the plant management be permitted to decide the cash factor in every special case with reference to the special conditions in question, to the degree of difficulty of the job, the worker's earlier earnings, etc.

The discussion about the problem, which was naturally regarded from the employer's point of view, developed in turn two main premises. One was that "an ideal state of affairs" exists, which meant that "the piece wage offered is accepted forthwith", the other that "conditions are as they really are in practice", that is to say, disputes occurred.

If in line with the first premise a fixed cash factor was used this had the advantage that large increases in earning could be avoided and the demands on the individual job study man were small. On the other hand the disadvantages were that less efficient workers, those who earned less, were favoured at the expense of the more skilled, which had the result that the workers became uninterested in continued rationalization.

If, using the same main premise, there was a variable cash factor that the job study man was able to vary—possibly within stated limits—earnings would be assessed by him according to the kind of work and skill of the worker so that the dislike of rationalization was counteracted. On the other hand there was the risk of a more rapid increase in pay and the demands on the time and motion study men became greater.

If the main premise was now changed and reality was calculated as it was, with disputes, the current agreement and decision of the Labour Court, what would the position then be?

If the cash factor was fixed, the earnings paid and the cash factor should in the short-term agree quite well, but in the longer-term the differences should increase. With increased practice and skill the worker should earn more on a piece wage, he would build it up. Then in the setting of a new piece wage a dispute was certain to arise if the old fixed cash factor was utilized at the same time as the worker's actual earning were disregarded. If instead the cash factor was variable, then the job study man and the factory management would be able to use a flexible variation so that disputes could be avoided. The job study then became "as of course was actually intended, only an aid in the setting of piece wages". In addition the workers would have an interest in promoting rationalization. On the other hand there remained the risk of pay increases that were much too high and demands on the job study man's abilities.[15]

Seen from the employer's point of view it was preferable that the cash factors were variable, as were other factors included in the piece wage calculation equation. In this way the system would be flexible but at the same time relative and dependent on the qualifications and judgment of the job study man. There was also the risk of uncontrolled increases in wages, the so-called wage drift. The opposite applied from the workers' point of view. With the cash factors fixed by agreement dependence on the discretion of the job study man's judgment decreased and the system became more secure. It more resembled the "objective" ideal, which job studies were still presented as, buth which in practice they never were. The criticism against the time and motion study men and their discretion permeates the arguments of the workers' party during the years around the mid 1940s.

Initially the aims of the Metal Trades Employers Association and the Metal Workers Union were thus diametrically opposite: one organization wanted variable bases for calculation, which were the company's own business, the other wanted fixed cash factors, preferably regulated by central agreements. But soon the development was to set both parties on the same tack: regulation of the cash factor was on its way.

The reason for the employers' turnabout, which had not really occurred at the discussion on principles in 1943 but which was to be noticed already in the following year in the arguments presented to the Labour Court and was about to find striking expression in the reactions to the draft agreement in 1947, is really quite simple. The risks of variable cash factors were that management was compelled to rely on the job study men and wages had a tendency to rise. As time went on this assumption of risk seemed much too large for the Metal Trades Employers Association. Mistrust of the job study men is well documented. Wage increases for piece wages outside the agreement were in the association's opinion unwarrantably steep.[16] Against the background of these assessments it was compelled to accept the idea of a fixed cash factor, a system that soon showed itself to have clear advantages in a negotiating situation, exactly as Styrman had visualized as early as 1943.

But at the negotiations of 1943 the parties also disagreed on the terms used: cash factor, grounds for calculation of earnings-factor. As Oscar Westerlund suggested in his report to the members of the policy conference in December, there lay in these differences of opinion an issue of a much more serious kind than the battle about terms suggested. What it was all about was the meaning of the term and also the right to the earnings that lay in the building up of piece wages. Did the guaranteed individual earnings-level as confirmed by the Labour Court in 1934 mean that in the setting of new piece wages one should proceed from the originally calculated earnings capability of the individual piece wage or from the individual worker's earnings after building up, that is to say when he increased his output through experience and skill. Was it in other words the earnings' possibility of the piece wage that was guaranteed or was it the individual worker's actual possibilities of income? The matter was brought to a head by submission to the Labour Court in the spring of 1944 when the Metal Workers Union brought an action against the Metal Trades Employers Association and STAL (Svenska Turbinfabriks AB Ljungström) of Finspång.[17]

In presenting its case the Metal Workers Union proceeded from a definite case. For a number of years a worker had been employed at a piece wage on a grinding job. The average earnings had been 1.83 Skr (Swedish crowns) per hour. At the beginning of 1943 the operation had been changed. Because a thicker working material was used the grinding time could be reduced so the management had requested that the piece wage should be changed. In order to get some foundation for this the work was subjected to job studies, which later resulted in an offer of a unit price of 4.02 Skr against the earlier 7.80. The worker had not accepted the unit price but, in spite of this, had started to use the piece wage rather than work at the hourly rate. During the course of 1944 it had been observed that the piece wage offered had been set too low. Earnings at the piece wage had varied between 1.28 and 1.53 Skr per hour, and thus had never come anywhere near the earlier average earnings of 1.83 Skr per hour. The Metal Workers Union now asked that the Labour Court should declare that the reduction of the piece wage was contrary to the current agreement with support of the interpretation regarding the individual earnings-level the court had confirmed in its judgment 1934/181.

The parties on the employers' side proceeded from a principle argument about the setting of piece wages in job studies and presented the standard formula for the setting of piece wages: $A = T \times U \times (1 \times S) \times (P/60)$ (cf. p. 42).

The other side, said the employers, now wishes to establish that the actual earnings-level, that is to say the earning since the piece wage has been built up, should be guaranteed by virtue of judgment 1934/181. This would mean that if a piece wage was set with regard to the earnings of 1.50 Skr (the money factor was set at 1.50 Skr) and thereafter built up to an actual earning of 1.60 Skr, so at a new setting of the piece wage the latter sum should be guaranteed, that is to say the money factor should be changed to 1.60 Skr. It should then presumably be possible to build up this piece wage to 1.70 Skr, which in its turn should be the money factor in the next setting of the piece wage. The result of this, said the association, would not be the guarantee of the status quo with regard to wages that the Labour Court had intended to guarantee, but a guarantee for permanently increasing piece wage earnings. The association considered that by guaranteed earnings-level must be meant instead that the employers were not entitled to use a lower cash factor than that valid when the agreement came into force. Finally, said the association, in the actual case "a tacit agreement" about the cash factor had existed for a long time at the company concerned.

The Metal Workers Union maintained its point of view and rejected the arguments in principle of the Metal Trades Employers Association. An "arbitrary" cash factor decided by the company could not be accepted as evidence that a piece wage was calculated in accordance with the agreement. In addition, according to the union, it was quite in order for a worker to be entitled to credit himself with the building up of a piece wage when a new setting of this was made. Otherwise the employer would discount the experience of the work obtained by the worker. The most reasonable method for the setting of a piece wage in the present case where the work was almost identical but the time required was less would have been to set the new piece wage in the same relation to the old, as the new time stood in relation to the old, which had given a unit price of 4.86 Skr instead of the 4.02 Skr offered. Even though the employer maintained that the old piece wage had been wrongly set it was impossible to avoid the actual earnings-level as the basis. It was only, said the union in a new interpretation of judgment 1934/181, when a wrong setting of a piece wage could be shown to depend on the workers' actions, that the employer had the right to alter it.

The last argument was rejected at once by the employers' side; the right to alter erronous piece wages was not affected by the reasons for the incorrect price-setting. The association also argued, against the argument that the employers discounted the workers' experience in the new setting of piece wages, that experience—to the degree it could be distinguished from skill—should be given weight in the performance index. In this way the source of error claimed should be eliminated. In the actual case concerned here the company had certainly not calculated with any performance index (which is equivalent to U=1, that is to say, normal assessment).

The Labour Court opened its judgment by emphasizing that it had already pointed out in its 1934 judgment that the individual case could hardly be adjudicated. It was first if an attempt could be observed in an employer to lower the piece rate contrary to the agreement in a series of cases that the matter could be taken up for adjudication. In the case before the court it was therefore a question of whether the actual case stated could be regarded as evidence of such an attempt. But in this case the parties also presented in their pleadings fundamental arguments on the interpretation of the 1934 judgment, so that the court considered it "should not avoid making a statement concerning it".[18]

This statement of principle was to mean in every point a recognition of the stand taken by the employers. With regard to the right to alter wrongly-set piece wages the court stated:

> With regard to the extraordinary difficulties associated with the setting of piece wages and to the impossibilities of permitting every piece wage setting ot be preceeded by careful time and motion studies, the Labour Court considers that it follows quite naturally that the correction of error in the setting of piece wages should be permitted to be carried out independent of the reason for the error in the setting and that the employers' responsibility to see to the maintenance of the earnings-level does not apply in such cases. Whether the erroneous setting has occured or not must naturally always be decided from objective premises.[19]

With regard to the method for the adjustment of piece wages the system proposed by the workers' side, which was constructed on proportional time, was rejected. This created a tendency to a continuous rise in the piece wage, which was much too dependent on the individual skills of the worker subjected to job studies and left no opportunity to utilize time and motion studies efficiently. The court continued:

> For the rational utilization of time and motion studies the employer is entitled to use as a basis for his piece wage proposal the method that has been described from the employers' side. But in accordance with this method it is necessary for a certain cash factor to be included in the equation. It is, however, quite obvious that this cash factor must be kept constant and may not undergo other changes downwards than it is possible to support by agreements arrived at concerning general norms for the wage level or otherwise dependent on mutual collective agreements.[20]

Following this the court spent some time dealing in detail with the problem of how the experience and skill of those who were subjected to job studies should be weighted in the equation for the setting of the piece wage price. In both cases the line of argument resulted in an emphasis on the importance of the performance index.

In the actual case—STAL of Finspång—the court considered it did not have evidence to be able to find an attempt contrary to the agreement in the management of the company.[21]

The court's decision, 1944/51, thus means that its line from the 1930s encouraging rationalization in these questions was pursued. The high estimation of the tendency of rationalization ideas to seek to make objective the antagonisms between the parties through scientific procedures became still clearer. It was in this spirit that the Metal Workers Union's more mechanical method of calculating new piece wages was rejected. The judgment also came to constitute a powerful impluse to follow the recommendation, which the agreement already implied, to set and legitimize cash factors through agreement.

3. The 1945 strike of the Metal Workers Union

When the Labour Court handed down the STAL decision it was not long before the preliminaries were in full swing for the great metal trades conflict of 1945.[22] At the ordinary Congress of the Metal Workers Union at the beginning of September 1944 there were four motions about rationalization and job studies.[23] One calling for further and more easily understood information to the members was approved by the executive board and accepted by Congress in the shape of a general instruction to the board for investigation.[24] On the other hand, two that wanted the government and the central organizations to take part in a joint institution for the provision of objective job study men were rejected for practical reasons.[25] The demand was to return in the Riksdag (parliament) motions in the spring (cf. p. 255). Finally the demand was advanced from Branch 4 in Malmö that an investigation should be made in order to place before the approaching policy conference the lines for the struggle against the darker sides of rationalization.

An unmistakable mistrust against the executive board came to light. The motion from Malmö was also aimed at the board, which with regard to paragraph 4 apparently intended to proceed only with certain completions of the demand of 1943. The aim was even more pointed as the motion required that the investigation should be carried out by people "working in production". The board moved that the motion should be rejected and maintained that no investigation was necessary. It would not get any further than that of 1938 urged Erik Collin in the debate.

Against Roland Tyrell defending the main line of the motion stood Lennart Eckerström, member of the executive board, who replied in a manner that helps to reveal the kind of antagonism that existed whilst at the same time illuminating the stand on grounds of principle taken by the board:

> ... Even in a socialist society we must count on a continued process of rationalization. Without this we cannot improve our real wages in the manner we desire. One of the causes of dissatisfaction with rationalization has been that some job study men have not been objective. You should not believe that the executive board regards job studies as a real science that solves all questions relating to piece wages in a painless way, but we have accepted them because this system often gives a fairer result than the old method of setting piece wages ...[26]

There were differences between a more radical line, which the communists and those of the unaffilated trade unionists joined, and the social democratic view of the board that once more increasingly clearly found expression. Again this was to set its stamp on the work within and with the union for many years to come. But here the Congress majority was still in the hands of the executive board whose proposal was approved by Congress.

Things were to be different some weeks later when the union policy conference on the agreement was held. Without any great discussion it approved the board's proposal to give notice of termination of the current agreement. The proposal for the new agreement gave priority to an increase in minimum wages. Piece wages were touched on in the chairman's presentation but on this point the board considered it was unable to propose any changes as the current provisions were so new. Nevertheless during the collective negotiations a new discussion would be initiated with the employers about the problems associated with piece wages. If this soft approach was not successful the union would prepare a completely new proposal for the piece wage paragraph. Such a fundamental overhaul would take a long time, however, so that it could not be utilized in the approaching collective negotiation activities.[27]

The board's recommendation for caution was not to the taste of the majority present. The latent dissatisfaction with the compromises in the previous agreement rose to the surface. The well-known demands were aired again: the rights to take the initiative, to receive reports, to obtain urgent agreement on the grounds for calculation, that representatives of the workshop clubs should always have the right to be present at job studies. In spite of opposition, especially from Erik Collin, the conference decided that the previous year's proposal for paragraph 4 clause 4 should be presented again in its entirety. Also noted in the minutes was the request by the communists Sture Norberg and Roland Tyrell from Stockholm about the possibilities of enforcing general methods for job studies throughout the country, and the right to negotiate on the grounds for calculation. These can more be regarded as demonstrations of strength.

Even in other respects the decision of the conference went against the executive board, for example in the election of a negotiating team where a number of new names supported a more radical emphasis. It seems that the board lost some of its control over the policy conference and no final proposal for a new agreement was accepted. On the suggestion of Westerlund the negotiating team was given authority to edit the proposal finally.[28] Perhaps it was to be easier to put matters right with a small circle involved.

The first meeting with the negotiating team was held on 28 September.[29] Erik Collin presented the proposal approved by the board and did not refrain from pointing out the lack of consistency in the demands of Tyrell and Norberg. The chairman of the Union of Foundry Workers, Sven Ekström, who took part in the discussions, noted that his union was not too happy about the freeezing of the grounds for calculation entailed by the proposal before them. Oscar Westerlund left the team in no doubt about his attitude to the decision of the policy conference:

> There's not the slightest chance of getting a proposal like that accepted. At the last collective negotiations we discussed this piece wage provision for a month and it was the subject of many deliberations, discussions and drafts, but nevertheless no better result was achieved. In my opinion if we do not now obtain adjustment, next year we shall have to delete this provision and prepare an entirely new one for the setting of piece wages.

Erik Collin detailed his criticism of the two statements included in the minutes of the policy conference. With regard to Roland Tyrell's he considered that the purpose was already covered by the proposal before them and that Sture Norberg's was simply a shot in the dark: almost all companies used the average value method in order to determinate the time selected and because of the STAL judgment there was no possibility of avoiding the utilization of the performance index. There was really no chance of variation in the methods. The delegation approved Collin's proposal that the statements entered in the minutes should be used during the negotiations but should not be added to the proposed agreement from the workers' side.[30]

The Metal Workers Union gave notice of termination of the agreement in due time and presented a draft where paragraph 4 was unchanged with the exception of clause 4, sections 2 and 3, and the entries in the minutes under the same clause. It was proposed that these parts should be deleted and replaced by the proposal of 1943.[31]

Within the Metal Trades Employers Association the negotiations were awaited with confidence. Notice of termination in the autumn of 1944 would come at a comparatively favourable time.[32] The associations counter-proposal was delivered at the beginning of December and met, on the piece wage question, a demand for a return to the wording of clause 4 that had been valid in the agreements from 1938 to 1943. Apparently this was a case of a demonstration policy on the part of the employers. Styrman writes the following in his history:

> The Metal Trades Employers Association at the beginning of the negotiations had advanced certain old wishes concerning the general provisions, perhaps essentially in order to 'air' these a little.[33]

It was this attack on the gains in the job study question during the previous year that specially upset the opponents.[34]

During the five separate rounds of negotiations held between the beginning of September and the beginning of February the parties did not come any closer on the important points in spite of the efforts by an appointed conciliation commission. The piece wage paragraph does not seem to have played any great part in the negotiations during this period. On both sides the demands about paragraph 4 had the air of a tactical feint.

In the negotiating teams the piece wage question did not stand foremost: minimum pay dominated the struggle. Certainly Oscar Westerlund retained his pessimistic evaluation of the chances of getting anywhere with demands about piece wages and was prepared to sacrifice these. On the other hand the majority line held to its view that the piece wage question together with that of the lowest paid were indispensable fundamental demands.[35]

On 26 January the Metal Workers Union gave notice of its intention to declare a general strike of its members in the metal trades. In spite of the efforts of the conciliation commision, which seemed to be becoming increasingly hopeless, the threat was put into effect on 5 February. The great strike of the metal workers became a fact. It was to become the largest conflict of the labour market in Sweden between 1909 and 1980. The hostility prevailing was so severe that no real initiatives were instituted by the conciliators. A feeler was put out in the middle of March without result and a further attempt was made a month later. The latter attempt at conciliation is of some interest as the piece wage paragraph was then introduced into the discussion.

On 18 April the conciliation commission presented a proposal that was discussed by the union's negotiating team but found to be unacceptable. In this situation Oscar Westerlund scented a tactical advantage in introducing the piece wage question; to press the demand for changes there in order to get the employers to come to an understanding on the wage question. Roland Tyrell, perhaps the leading representative of the communist line, preferred other tactics: initially no linkage between minimum pay and the piece wage question.[36]

Once again the negotiating team's discussion on the piece wage question confirmed the almost insuperable difficulties of discovering a satisfactory construction for the draft agreement. Should the provision be worded too carefully it might be found to counteract its intention. Instead it might be better to try to develop fundamental principles that could be used by the parties from case to case to obtain agreement.[37]

The procedure recommended by Tyrell was followed. The workers' side presented a wage offer including an increase in the minimum wages of 11 öre per hour. The employers made a counteroffer of 6 öre, which was rejected. The conciliators then made it clear that it was again prepared to break off conciliation attempts. Then the workers' side played the piece wage card. It was prepared to accept a wage bid equivalent to the agreements made in other unions on condition that an acceptable counterproposal was received in the piece wage question. At this stage, of course, the position was that the workers had demanded a construction they had been unable to obtain the previous year whilst their opponents demanded a return to the prewar position. What was needed now was a constructive offer that could be used as a basis for negotiations.

In this position it was natural for the conciliation commission to again follow the line it had adopted in the previous agreement, that is to say only to modify the agreement. In this it had apparently been aware of the workers' assessment.

The conciliation commission's proposal of 20 April meant that a new addition should be made to clause 4. This would regulate the setting up of a joint committee "for the furthering of confident cooperation and good working conditions and for the settlement of differences in everything concerning job studies and the setting of piece wages". The committee should be able to investigate stated anomalies at workplaces and provide instructions for their solution. If these were not followed negotiations in accordance with the negotiation procedures could be requested. The committee was thus not to have the right of decision.[38]

This proposal links up with the discussion in the Metal Workers Union's negotiating team to the effect that instead of further regulation in detail of the agreement an organ should be created for the interpretation of questions of principle. But this line of thought had a longer history. As early as the policy conference on the agreement in the autumn of 1944 Erik Collin had noted the difficulties of arriving at a wording of the agreement "that prevents the employers' desire to cut". Agreements on the grounds for calculation were recommended, but what would happen if such an agreement could not be reached? Collin provides the answer himself:

> So far as I can understand there is hardly any other solution during the period of agreement but to accept an arbitration tribunal procedure. As we have discussed the matter in the executive board we have come to understand that the employers' side has no objection to such a procedure and for this reason we don't consider we should present a proposal in this direction.
> Following careful discussions we have come to the conclusion that the most suitable method of procedure from our side is not to make any proposal . . .[39]

A combination of agreements concerning grounds for calculation and an organ for arbitration was regarded by the representatives, even before the pay discussions had started, as the most sensible solution. The conciliation commission was also to follow this line.

Collin's statement also illustrates an aspect of trade union tactics, which was based on an idea of what the opponent wants. Nor was the employer's side negatively inclined to an organization for arbitration on piece wage questions. It is also in line with the spirit of the General Agreement of the main organizations (the so-called Saltsjöbaden spirit) and the working methods of the Labour Market Committee. In addition it is of interest to recall the proposal put forward by Uno Forsberg in the general council of the Metal Trades Employers Association in conjunction with the collective negotiations of 1931. (See pp. 66 et seq.). Even in a third instance, which proved to be of great importance in the earlier development, the proposal seemed to be well-established. In its judgment on the STAL case the Labour Court, when the right to alter faulty piece wages had been confirmed, used the following words:

> Whether the wrong setting has occurred or not must of course always be assessed on objective grounds.

The significance of this wording must be that an arbitration procedure, not negotiation, could resolve the kind of disputes associated with piece wages. This implies in its turn the assumption that an organ for such arbitration can be set up on the basis of norm constructions acceptable to both parties. This construction of norms was to be found in the methods of job studies. The proposal on the setting up of a job study committee would thus be associated with a view that had existed earlier, at the same time as it fitted in with the course of negotiations quite smoothly. After having conferred with the Trade Union Confederation the workers' side was prepared to accept the conciliation commission's proposal. But there was great dissatisfaction because the committee had not been provided with decision functions.[40]

However, the employers totally rejected the conciliator's offer, mainly because the workers' side was unable to state its pay claims exactly. Neither were they prepared to discuss horse-trading on the piece wage question on such a feeble basis.[41] The idea of a job study committtee was therefore not subjected to any real assessment by the Metal Trades Employers Association's negotiating team.

Again the negotiations broke down on the pay question. On 7 May the conciliators made another attempt to get the workers to relax their rigid stand. The trade union functionaries also tried in every way to influence the majority of their negotiating team but without result. The so-called 11-öre line in the pay question stood fast. At the suggestion of Oscar Westerlund the policy conference was again reconvened.[42]

The chairman reported on the position following the negotiations of 18–21 April, explained the employers' view on the piece wage question, presented the proposal of the conciliation commission about the job study committee as well as making clear the position taken up by the workers' side. Westerlund said the negotiating team had stood firm on the 11-öre line and thus rejected the idea of a job study committee. That this was contrary to his own and the executive board's assessment of what should have been done was unmistakable. The board's recommendation for continued action meant that the union should adopt an 8-öre line and at the same time try to obtain improvements in the piece wage paragraph. In the last resort it should be possible to accept the proposed committee.[43]

Opposing this line was the tougher, communist one, which thought that the workers' side should maintain its demands in both the minimum wage and piece wage questions. That this should be feasible was backed up by the communists' assessment of future trade conditions. They considered that the recent German capitulation in Europe should bring about a rapid boom that the employers were not prepared to sacrifice. Spokesmen for the general council opposed this prognosis and instead feared a depression. The communist proposal was carried by one vote.[44]

The conciliation commission's proposal of 12 May, suggesting a 6 öre increase in minimum wages could thus not be accepted by the union's negotiating team, even though a further alteration was made in the piece wage paragraph. Minute 3 of clause 4, which included a recommendation to make agreements on the grounds for calculation, it was proposed, should have a more binding formulation: "At companies ... the employers and employees should make such agreements".[45]

In this position the commission decided, after conferring with Sven Ekström of the Foundry Workers Union among others, that the only possible chance was to present a proposal over the heads of both parties. This was done and on 14 May a draft agreement was presented. The executive board of the Metal Workers Union could not refuse to ballot its members on this. The returns were in by the end of the month and resulted in a clear negative to the proposal.[46]

At the discussions held between the conciliation commission and the union when the decision was presented, the commission offered to make a further attempt at conciliation on condition that the employees' side made a definite realistic wage bid. But the negotiating team was unable to agree on this and negotiations were broken off. The chairman of the conciliation commission reported that no new attempts would be made until the employees' side so requested, or the employers accepted the 11-öre line. In this situation the general council of the union, which contined to recommend the 8-öre line, sent an informatory circular to the members, also with the idea of feeling out current opinion. This showed that the mood within the union was moving in favour of an agreement. Therefore the policy conference reconvened on 18 June.[47]

The debate at this third conference was just as fierce as it had been earlier. The executive board proposed the 8-öre line and a piece wage paragraph in agreement with the conciliation commission's proposal of 12 May, that is more compulsive provisions for piece wage agreements on the basis of calculation and the setting up of a job study council. In addition an alteration was also proposed in minute 1 to the effect that the worker requesting the job study should not be compelled to give reasons for this.

The communist and unionist representatives of the more radical policy held in principle to the decision of the previous conferences, but in the circumstances were also compelled to use hypothetical arguments. If the conference accepted a lower wage-level, said Knuth Svensson of Stockholm and others, then this concession must be compensated by keeping to the original demands in the piece wage question. Oscar Westerlund agreed that improvements should be sought on this point but it might be difficult enough with the limited alteration suggested by the board. In fact Erik Collin was compelled by the criticism of the piece wage question which followed to defend the existing provision and also to refer to other solutions than those offered by the collective negotiations:

> I should like to suggest that in the previous metal trades agreement, which expired at the end of the year, we even had the best job study provision that any collective agreement contains—in any case such agreements as are arrived at with an employers' organization belonging to the Employers Confederation ... For my part I believe that the attention that has been given during negotiations to time and

motion studies will result in conditions at workplaces becoming better. The Riksdag's decision on the question will also certainly have a beneficial effect. It is also intended that the parties of the labour market shall consider this question in the Labour Council.[48]

Collin also criticized the "immature job study men", who were the cause of a great deal of irritation. Axel Rosell from Trollhättan agreed on this point but not otherwise:

I don't think that through motions in the Riksdag or by any orders from higher levels we can get rid of the snotty-noses who get themselves qualifications by taking positions as job study men. It depends mainly on the management of companies whether we have to put up with them in the future or not.[49]

When it came to the vote the executive board's proposal was opposed by one that meant that the policy conference should keep to the earlier decisions. Again the result was achieved with the smallest possible majority, but this time in favour of the board.[50]

Negotiations before the conciliation commission were restarted on 19 June. Both parties accepted the 8-öre line and the negotiations also dealt with the piece wage paragraph's construction on the basis of the commission's proposal of 12 May. By 21 June it was possible to reach agreement on all points. In the piece wage question the conciliation proposal was accepted. Also minute 1 was altered in the way requested by the Metal Workers Union. A demand that workers who wanted job studies carried out should give special reasons for this was limited to those cases where the work had already been subjected to time and motion study. The conciliation proposal approved by the negotiating team was sent out to the union members for balloting. The executive board used all its authority to persuade the members to vote in the affirmative whilst the communist delegates demonstrated their negative attitude quite clearly. The ballot resulted in a defeat for the board: 37,908 votes against and 30,721 for; only 65.6% of the members voted.

In the performance that followed the piece wage question does not seem to have played any independent role. The workers' negotiating team, now greatly divided because of the events accompanying the ballot, asked for some further small concessions from the conciliation commission. However, the employers' ranks stood firm: the proposal should be accepted completely or the negotiations would break down again. If this happened the employers were prepared to extend the conflict with a lockout. It may be presumed that the iron and steel workers would be the first affected. Faced by this threat matters came to a head in the executive board. The unaffiliated trade union fraction yielded whilst the communists demanded that the government should interfere. As the negotiating team was still in disagreement the board by virtue of paragraph 14 of the union's standing orders decided to use its veto right and approve the proposal for agreement. In the early morning hours of 6 July the new agreement was signed and the battle was over.[52]

To extract and assess the importance of one factor from great events is of course always very risky. If the course of events is factually complicated, extending over a long period of time and marked by many varying rounds of negotiations the difficulties become much greater. To deal with these would require an extremely sophisticated model for analysis with the risks this would entail. The part played by the piece wage problem in the negotiations leading up to and including the metal workers' strike of 1945 will here only be made the subject of some observations.

What caused the greatest open differences of opinion was the question of raising minimum wages. It was around this the sparks flew between the negotiating parties and the two opinions in the negotiating team of the Metal Workers Union. The piece wage question with its regulation of job study procedures was given a tactical function from the beginning by all the parties. Both employers as well as workers went to the negotiation table with tactical overbids, or with the accent on clause 4. The former seemed to have aimed for the status quo, whilst the aim in the union, which later became that of the social democratic minority seems to have been set on following up the gains of 1943 by completion on certain points. The radical line in the union differed in its assessment of job studies from the executive board. It had no proposal of its own but demanded more generally the maintainance of the more radical program of 1943; it is difficult to avoid the impression that this was a demonstration policy on their part. The demonstration was perhaps initially directed against their internal opponents.

The tactical character of the demands in the piece wage question came to light when Oscar Westerlund wanted to bring this matter into the active negotiations in order to relieve pressure on the pay question. In a similar manner the communists tried to play tactics in June in order to cover the retreat that the changing attitude of the members seemed to make increasingly unavoidable.

So the tactical role to be played by the piece wage question should not be interpreted as a sign that the parties thought this problem was less important. There is a great deal of evidence that the opposite was in fact the case. Instead the doubts that existed about making the piece wage question the subject of real negotiations should be understood to mean that the parties were unsure how to proceed. This uncertainty had its basis in the fact that the material provisions of clause 4 were new and their effect were so far untried, even though the STAL judgment had gone a little way in that direction.

The result of the 1944–45 negotiations on the piece wage paragraph therefore became changes in details along the same lines the parties had already introduced at the previous collective negotiations. In addition there was the new job study committee. This actually changed neither the provisions nor the practice in connection with piece wages and job studies. It created a body that need not necessarily become a future burden because a committee could fulfil its arbitration tasks regardless of the material content given to the piece wage paragraph of the agreement. So it was not in the least dangerous to set up the committee even though it should not be regarded as a feint because of this. This marking of a common view as the basis for the activities that lay in the setting up of the council was a signal from both parties. From the struggle was born an organization for collaboration.

4. Collective negotiations 1946

The repercussions of the 1945 strike and the acute differences revealed in the Metal Workers Union were to last a long time. The air was cleared at an extraordinary Congress at the middle of February 1946. Soon after this the real preparations for the next collective negotiations were started. The agreement signed in July of the previous year was valid until the end of 1946.

At a policy conference in 1944 about the collective negotiations Oscar Westerlund had already stressed the need for a more comprehensive examination of the agreement's text on the piece wage question, and it was seen that the current wording was incapable of ensuring the workers' interests. During the whole of the ensuing collective negotiations the shifting assessments and lack of unity within the union on possible lines of advance were demonstrated time and again. The need to rally round the program was clear. There was a meeting in 1946 of a committee appointed to prepare the collective negotiations and to examine carefully a proposal for revision to put before the union's policy conference to discuss the agreement.

Initial discussions in this committee produced some idea of the conditions considered to be especially unfair whilst proposals for definite solutions were, as always, difficult to discover. This part of the planning was handed over to a subcommittee. One of the most important questions was still about guarantees that earnings should not be allowed to fail. One idea was to obtain a provision similar to that introduced into the iron and steel trades agreement in 1937, which allowed a trial period of a month for every piece wage. But many were sceptical about it; this arrangement was all right in the iron and steel mills but in the metal trades' plants, where most of the piece wages were temporary, it could not be utilized. So there remained instead to obtain such influence and control over the setting of piece wages that it was possible to assess the piece wage offered. The local trade union organizations had to be guaranteed an absolute right to negotiate on the grounds of calculation and the bonus for lost time; but to introduce general grounds for calculation into central agreements was something that many members of the committee were doubtful about. Then there was the problem of obtaining control over the actual job study situation on the factory floor. To get rid of the potential arbitrariness existing in the performance index was impossible following the attitude of the Labour Court in 1944. Therefore the right of the local unions to be represented at job studies to support the individual member became much more important.[53]

The draft agreement was drawn up in the subcommittee headed by the representative Lennart Eckerström, but later on the executive board essentially made it its own.[54] This proposed that a worker who carried out a job where the piece wage could not be offered in advance because of job studies should in the meantime receive "his earlier average earnings received for piecework". The provision should mean that workers should not be compelled to go on hourly pay because of job studies. The wording was made somewhat more pointed in minutes 1 and 2.

Minute 3 also received a more compulsive wording. Negotiations on the grounds for calculation and lost time bonus were to be instituted locally and completed as soon as possible, but at the latest within two months. If no agreement was reached the matter was to be referred to central negotiations. Finally a further minute was suggested that should give the organization's representatives the right to copies of all such job study material on which the individual piece wage offer was based.

The committee for the inspection of the metal trades' agreement resulted overall in very extensive proposals for changes in it. In addition demands were proposed for large wage increases. But on the piece wage problem the committee was unable to present any revolutionary ideas, which had actually been hoped for, but it tried through more compulsive wording to strengthen the construction of the current agreement. Behind this choice of paths lay of course a realistic assessment of what was possible of achievement. The agreement was not considered to be worse than that it was possible to work further along the same lines.

The policy conference in September 1946 proceeded under the banner of the cautious offensive. The Trade Union Confederation's signals in the same direction were made initially by the union's chairman, who also reported that the trade situation was good and that, so far as it was possible to judge, the employers favoured a peaceful settlement. In other words the situation called for an assault, but:

> The collective negotiations may not be carried out in such a manner that we end up with a conflict. The metal trades' conflict of 1945 may not be repeated. We must utilize the machinery of negotiation in so far as it is possible. The union's finances do not permit an open clash.[55]

In spite of this Oscar Westerlund recommended that notice of termination of the agreement should be given, not because the system with preliminary negotiations had been unsuccessful but more as a concession to the members' opinions. Without any debate the conference approved his recommendation.[56]

The discussion was mainly about the new proposal for guaranteed earnings whilst awaiting time and motion studies, which if it succeeded would in Eckerström's opinion solve the greatest remaining question in dispute in paragraph 4. Some delegates queried whether the average earnings from piece wages should not apply until these piece wage rates had been *confirmed* by agreement. But in the opinion of the unanimous assurances from those who had formed the advisory committee such a formulation would not function in the manner intended. The executive board's proposal gave the piece wage workers the protection they could demand. The conference approved the proposal on paragraph 4 in its entirety.[57]

The Metal Trades Employers Association's internal analysis of the workers' demand on the piece wage question produced detailed arguments about the proposed guarantee of earnings in clause 3; the other proposals were brushed aside in a more routine way. The group carrying out the examination, the association's time and motion study committee, advised the general council to insist on maintaining the status quo.[58]

When the collective negotiations got under way the hostility between the parties, as expected, was very great, especially with regard to the pay demand. The tug-of-war was also accompanied by arguments on general economic policy including references to the price situation. In the negotiating team of the Metal Workers Union, there were the usual differences of opinion about tactics and the problem of the advantages of centralization.[59] The bitter struggle on pay questions put the demands for changes in the agreement's general provisions in the shade and in the chairman's opinion this increased the chances of the workers' side obtaining what they wanted.[60]

The negotiations were to carry on for a long time. The next stumbling-block was the pay question, but the delay during January and February also contributed to the wild protest action of workers at some plants. The executive board of the Metal Workers Union reconvened the policy conference at the beginning of February in order to inform the members about the position and to make it quite clear to the delegates that protests only injured the workers' cause instead of speeding up the negotiations. The discussions at the conference dealt with the tactics to be used and the wage level that could be demanded or expected. In this situation Westerlund expressed his optimism about the changes in the general provisions of the agreement. There was never any discussion about the piece wage question.[61]

In actual fact this question had already been removed from the line of battle. The compromise proposal, later to be included in the agreement, had already been presented by the conciliation commission at the end of January and these were the gains put at risk by the wild actions.[62] The Metal Trades Employers Association threatened to rescind the concessions made, said Westerlund at the same conference.

The new agreement was not concluded until 3 March 1947. It meant that two of the workers' demands for changes in the piece wage paragraph were obtained. The right to negotiate on the grounds for calculation was strengthened as the time limit requested was included in the agreement: if a local agreement was not concluded within two months, the matter could be referred to central negotiations. The main organizations were given the right to examine the time and motion study material in individual cases when there was a dispute. Demands for the rewording of minutes 1 and 2 were rejected, but these were of less importance. A greater reverse was that the earnings guarantee in clause 3 could not be realized. It was on this point that the employers' opposition had been mobilized to its maximum.

5. Collective negotiations 1947: money factors are regulated

The agreement of March 1947 was only valid until the end of the year. Therefore it was already necessary in the summer of that year to decide how the approaching collective negotiations were to be carried out.

The ordinary Congress of the workers' union, held 7–12 September, only dealt in passing with rationalization questions. A motion was put before Congress that reiterated the demand for information, which had already been dealt with at the previous Congress in 1944.[63]

In view of the unstable economic position and because it did not want to be first with this demand once again, the Metal Workers Union agreed with the Metal Trades Employers Association to delay the date for notice of termination of the agreement until 15 November. The union's policy conference to discuss the next agreement therefore did not assemble until the week before this date.[64] Without any great debate the conference approved the executive board's proposal on the piece wage paragraph. This meant that the previous year's demand for an earnings guarantee was repeated. It was demanded in a rider that special attention should be given to welding work in the setting of piece wages by time and motion studies. In addition in the draft for a new agreement that was handed to the Metal Trades Employers Association the following was also noted:

> In connection with the treatment of the piece wage paragraph, the workers' side whishes to discuss the grade structure for the determination of the basis for calculation of piece wages with regard to an eventual agreement about this and generally current money factors for the different cost-of-living index regions.[65]

This expression of the desire to lay down general grounds for calculation meant a change of attitude in the general council of the Metal Workers Union. The union had fought for the right to conclude local agreements for a long time. But at the policy conference in the autumn of 1944 Erik Collin, for example, had uttered a warning against locking the grounds for calculation centrally because this could affect those who received piece wages based on cash factors above the average.[66] Now, however, a couple of years experience of negotiations on the grounds of calculation had shown the need for central norms. So initially it was a grading structure for cash factors, that is to say how many and which categories of workers would be distinguished when cash factors were established. Some employers reckoned with as many as 14 groups. Against this the union's general council wanted to establish the cash factor for four groups and five cost-of-living index regions. On this point there was concord with the Metal Trades Employers Association.

Behind this rethinking of the union lay undoubtedly not only the experience of the union's officials but also complaints from union branches and works clubs. Note the impact in a communication from the works club of Eccoverken at Skara:

Like all things human even time and motion studies are an idea for good and evil. With arbitrarily established local grounds for calculation in the setting of piece wages the employer takes back what the worker has perhaps won at central wage agreements. The pressure of the work will soon exceed human capabilities. The quality of man and the product must become fundamental in human activities even at the cost of one or so tenths of a second. If we are to be able to fulfill our historic task in the service of social progress we may not sacrifice every spark of life on the altar of the production process.[67]

The policy conference decided to follow the general council's recommendation not to terminate the agreement but to try to force the new proposal through at the preliminary negotiations.[68]

The initial contacts were made at the beginning of December but the first real negotiations started in the week before Christmas. Subcommittees were then appointed for different matters, including one to discuss cash factors and the grounds for the setting of piece wages.

Prior to this the Metal Trades Employers Association had gone through the demands of the other side internally. The association's technical committee based its discussions on a memorandum written in connection with the initial round of negotiations:

Memorandum concerning grounds for the calculation of piece wages.
Confidential
At the collective negotiations on the agreement the workers' negotiating team expressed the wish to have detailed provisions inserted into the agreement about the calculation of piece wages. This is perhaps an opportunity that will not recur. It would seem to be very difficult for the Metal Trades Employers Association, when the employers themselves are compelled to raise the question as a necessary result of the development, to reach such an agreement. It would appear that the development because of the 'recommendations', which have been successfully introduced in clauses 3 and 4 of paragraph 4 of the agreement, results in a central direction of the agreements on the grounds for calculation of piece wages at the different plants.

. . . it has appeared that time and motion studies and the setting of piece wages at a plant that has obtained an agreement on the grounds for piece wage calculation proceed considerably better than before and from various quarters wishes have therefore been expressed that the Metal Trades Employers Association should try to arrange the question for all plants through detailed provisions in the agreement . . .

If both the grade structure as well as the cash factors are fixed in the agreement irritation and envy etc. are avoided between both members as well as workers . . . If the grounds for calculation are fixed for the whole of the metal trades industry it will be easier to control that time and motion studies are dealt with correctly (which is

> necessary if serious consequences are not to arise soon) than if each plant has the possibility to 'fiddle' (tendencies in this direction have been observed in many quarters). If the cash factors are fixed in the agreement they will also be more stable as the Metal Trades Employers Association has of course greater possibilities of taking care of the question in the above manner at the negotiations on the collective agreement than the individual employer has when the workers come with demands for increases. . . .[69]

Attached to the memorandum is a draft of how the provisions of the agreement on cash factors could be drawn up. These coincide essentially—but not in the amounts stated—with those that later were written into the agreement. The discussion in the technical committee revealed no differences in the assessment. The Metal Trades Employers Association now wanted central agreements on the grounds of calculation, also because of the oppressive negotiations that had been carried out because of minute 3. This development showed that Georg Styrman was correct in his assessment of 1943: from the operational point of view fixed factors were an advantage. The Metal Workers Union's demand for an earnings guarantee during time and motion studies was still rejected categorically as was the proposed time bonus in welding.[70]

The subcommittee working on the question of the cash factor had made a good start and was soon in agreement on everything concerning the wording whilst questions about top figures and coverage between groups and regions were not completely clear when the negotiations were resumed at the beginning of January.[71] This meant that once again the piece wage problem was got out of the way at an early stage; the demand about an earnings guarantee does not seem to have been discussed seriously at all. The collective negotiations were dominated by the question of minimum wages as was the case the previous year. The negotiations, with Olle Ekblom as neutral chairman, broke down on this question and a policy conference was convened by the Metal Workers Union for 14 January to provide information.

A conciliation commission was set up later and was able to present a proposal on 29 January that was accepted by the negotiating teams of both parties. After polling of the trade union members the new agreement was signed on 16 February 1948.

Thus the agreement of 1948 contained a new clause under paragraph 4. This, inserted as clause 5, laid down by agreement the formula for the calculation of piece wages in connection with time and motion studies which had been used unilaterally by the employers for many years and also had received the sanction of the Labour Court in the STAL judgment of 1944. The formula was somewhat modified compared to the one presented by the Metal Trades Employers Association at that time. Because of the Labour Court's view, experience was included in the calculation. The performance index was stated on the basis of an average skilled worker. Cash factors were stated in öre for four different grades of workers and five different cost-of-living index regions. Cash factors also became of importance in cases not subjected to time and motion studies. They were then to function as "targets in the normal working performance of an average skilled worker".

Some provisions, partly of a transitory character, were also added as minutes. These emphasized that the new clause did not alter the validity of clauses 2 and 3. The long-maintained demand of the workers' side about protection against decreased earnings had already been rejected by the conciliation commission, which had given a time limit for when the old grounds of calculation for piece wages should cease to apply in the new setting of piece wages (6 weeks) and scope was provided for the special treatment of exceptional cases. Finally it was confirmed that the new provisions would not be applied retroactively so that they could constitute the basis of demands for the alteration of existing piece wages.

It seems as though both parties were satisfied that the piece wage paragraph had been given a more precise construction. The preliminary assessment fo the Metal Trades Employers Association has already been shown. At the meeting of its general council in August 1948, Styrman reported that experience of the fixed cash factors had so far been good. A discussion arose around the question of whether minute 1, in the interpretation given by the Labour Court in its judgments of 1932, 1934 and 1944, still applied. Styrman maintained that as the cash factors had been laid down by agreement the minute had ceased to apply to temporary piece wages.[72] The Metal Trades Employers Association had let their opponents hoist themselves with their own petard. The introduction of cash factors brought about by the Metal Workers Union had destroyed the basis of the union's daily wage struggle that had been carried on for a quarter of a century. Naturally it felt like a victory. But Styrman was counting his chickens before they were hatched. This interpretation was tested later by the Labour Court in case 1953/44, when the court declared that the minute must still be considered to apply in the interpretation given successively in 1932, 1934 and 1944. On the workers' side also the negotiators were satisfied, at least to begin with before the disputes about interpretation had got under way. Back summarizes Westerlund's reactions in the following manner:

> In another commentary Westerlund emphasized his satisfaction that the union had at last succeeded in getting their opponents to agree to a provision concerning the calculation of piece wages that generally could constitute a firmer ground for the fixing of piece wages than the previous rules, which meant that there had been extremely different grounds for calculation at different workplaces.[73]

So true enough, both parties thought they had led their opponents by the nose and got the best of the deal. At the same time it is important to see the agreement of 1948 as a milestone in the development of a compromise in the field of rationalization. Both parties agreed about a technical method for the calculation of piece wages. A rational division was finally acknowledged between a more objective factor—time measurement—and one that was the subject of negotiation—the cash factor.

6. The Job Study Committee and disputes about piece wage questions

The time and motion study committee, which was instituted through the agreement of 1945, was no hard-working organization. No real archives of the committee have been discovered but it appears as though it only had two cases to deal with during the 1940s: Ystamaskiner on 3–4 January 1946 and Svenska Fläktfabriken of Jönköping on 28–29 May in the same year.[74] It seems safe to suggest that the committee's most important function in the short term was to constitute a message of compromise in the metal trades conflict and in the long term to constitute a prototype for the approaching time and motion study agreement between the Employers Confederation and the Trade Union Confederation.

One reason for the time and motion study committee's inactivity lay in the formulation of its tasks, another in the frequency of disputes and disturbances around time and motion studies. It did not replace the central negotiations in accordance with the procedure prescribed by the metal trades agreement. Disputes concerning eventual earnings were dealt with by the procedures laid down in the agreement. Disputes about the methods and execution of time and motion studies became increasingly rare.

One source that can be utilized to obtain an understanding of the number, frequency and signficance of disputes is the organizations' dispute registers. The possibilities of these have not been exhausted by any means in connection with this study. A few observations only will be given here.

The dispute register of the Metal Workers Union, which for practical reasons is a good point from which to proceed, consists of a card index where every dispute carried to central negotiations is represented. The register is arranged according to the paragraph in the agreement to which the dispute applies, the branch number and companies in alphabetical order, as well as the date of the dispute. Every card carries these data together with an account of the dispute and the negotiations undertaken. The cards provide information about the documents preserved about the disputes. Altogether 1,800 disputes about paragraph 4 of the metal trades agreement are shown. In less than 10% of these reference has been made in some connection to time or work studies. About half of the latter apply to demands about control studies. From approximately 80 disputes is shown more generally, "dissatisfaction with time studies", "incorrect time studies", "dissatisfaction with time study man", etc.[75] The number of disputes was thus, regarded absolutely, comparatively small. Most of these were resolved at central negotiations.

More striking and more important in the formation of opinion, at least outside the circles most closely affected, was a small number of disputes or incidents that led to strikes or inappropriate behaviour. Totally the union's register shows 9 cases of work-stoppages during the 1930–40s, 9 cases of inappropriate behaviour from the workers' side, which in some cases was combined with a strike, and one case headed "Various".

The Metal Trades Employers Association's register includes 9 cases of conflicts, of which 3 are not to be found in the union's register. In addition there are 2 cases under the heading "Various", neither of which corresponds to the one shown in the union's register. Together the registers of the union and the employers association provide a reason for investigating 19 cases. Of these 12 cases concerned dissatisfaction with the time and motion study man (engineer, chief) as a person, 2 concerned dissatisfaction with the time and motion studies as such, and 5 were about both the person and matter concerned. All of the conflicts except for one were solved at central negotiations, usually through some kind of change of job. Thus the extent of serious disturbances in working life because of time and motion studies was not great.

7. Summary

In this chapter the development of the piece wage paragraph of the metal trades agreements from 1943 to 1948 has been analysed. During this period both parties were more prepared than earlier to take up technically complicated negotiations on facts concerned with piece wages and time and motion studies.

The Metal Trades Employers Association presented a consistent opposition against the regulation by agreement of time and motion study activities, but from one collective negotiation to another the association was compelled to make concessions. In 1947 the attitude of the association swung from protecting local freedom of action to recommending central agreement provisions with fixed cash factors. Behind this change lay an assessment of a number of factors. Wage drift had shown itself to be a problem that might possibly be solved better with fixed grounds of calculation. The independent scope for action of the time and motion study men needed to be decreased to gain the procedural advantages of the simple rules. The problems of time and motion studies in the calculation of piece wages thus compelled the association to extend its field of action and intensify its contacts with the other side.

For the Metal Workers Union it was a case of securing the gains made by the workers through negotiations or otherwise, with regard to wages. One of the main tasks was to limit the freedom of action and arbitrariness of the employers, and the job study men, by means of provisions in the agreement. The demand that the local organizations should get the right to make agreements on the grounds of calculation developed into a demand for centrally determined cash factors. In this way at least one of the variables in the equation for the calculation of piece wages was fixed. The central organization became of greater importance.

For both parties the Labour Court's judgment of 1944 meant that limits were drawn for the possibilities open to union policies. Lines of development were projected from which it was not worth while diverging too far. The court laid down norms for a changing practice.

Analysis of the part played by the piece wage question during the metal trades conflict in 1945 has shown that for both parties it was initially of tactical importance. This may be interpreted to imply that no clear radical alternatives had been found by either side to the construction of the piece wage and time and motion study provisions received in the agreement of 1943. Following the conflict the time and motion study committee was established. However, this became of little importance other than as a pattern for later central agreements because disputes about the principles of piece wage setting and methods became fewer. Disputes about the actual calculations were taken up for central negotiation in the usual manner.

CHAPTER 6

Piece wages and job studies in the metal trades agreement

1. Summing up

The course of events already analyzed here covered two decades, 1928–48. During this period job studies in the metal trades industry began to spread to such an extent that the central organization of the Metal Workers Union was faced by the demand to state its attitude to them. For an increasing number of members these studies seemed to be a problem. There was a growth in irritation about the studies.

The union converted this annoyance into demands for changes in the text or application of the metal trades agreement. The Metal Trades Employers Association opposed so long as it could a regulation of collective central agreement on the piece wage question and the application of time and motion studies. The association thought that these areas were the concern of the individual company and to a certain extent the subject of agreement with individual workers.

What was most important for the workers in these problems was to maintain earnings in spite of the changes in the production process and in spite of the employers' opposition. The union tried to ensure security of wages by means of interpretations of the much discussed minute 1. With this end in view the union went to the Labour Court. In a series of judgments the court presented an interpretation of the significance of the minute during a period of industrial transformation. These successive formulations of the import of the minute rested on a construction that with increasing definiteness presumed that time and motion studies were used. At the same time the Labour Court's interpretation meant a defeat for the line the Metal Workers Union had urged to the mid-1930s; to include all piece wages in the earnings guarantee that it was thought lay in the minute. The principal difference between fixed and other piece wages was upheld by the court and the employers. So the union tried another tack. As the central setting of norms by the minute did not appear to satisfy the desired function the importance of local influence was therefore stressed instead. Fixed piece wages listed in schedules were declared to be the norm whereas the temporary piece wages over which the organization had no direct influence were described as exceptions.

This attempt to increase the influence of the local union organization thus found its first expression as in this way an attempt was made to undermine the free piece wage negotiations. The next step in the change of equilibrium was through demands for information about and influence on the grounds on which calculations of piece wages were made.

These interpretations of the agreement, together with the fact that job studies were also used increasingly for technical reasons, made it necessary for the Metal Workers Union to obtain information about job study methods and control over their actual application. Similarly the employers' association was compelled increasingly to take up the question. Central influence was once again emphasized.

The build up on both sides led to a hardening of antagonisms, which culminated in the metal trades strike of 1945. During the previous year, however, the agreement's piece wage paragraph had been rewritten. The strike produced no important change except that a committee was established that also dealt with arbitration problems. In practice the application of the agreement's provisions went well and during the last year of the agreement small adjustments were carried out and, with great unanimity, a regulation of the cash factors. Job studies as a semi-automatic and less disputed system were at last established.

In the national agreement norms were inserted on how these studies should be carried out. The cash factor, the aim of piece wage setting, was regulated in similar, centrally-binding forms to time wages. The union was assured a local influence.

In this light it appears obvious that the workers' union largely succeeded in neutralizing the threat to the balance between the two parties, which time and motion studies had appeared to be at their introduction. To a certain extent the union's countermeasures had even made it possible for the organization to advance its positions. Influence over the piece wage agreements and the grounds for the calculation of piece wages undoubtedly moved into an area that earlier had been principally the sphere of the employers: how production was to be organized.

Regarded from the employers' point of view it must be said that management and the employers' organization were unable to keep piece wage agreement and job study questions away from the area of activities of the union organizations. At almost every round of collective bargaining the setting of norms became increasingly detailed. The big question was how tenacious the employers' defensive lines were to be. A swing in the reactions of the employers' organization is noted in a few years into the 1940s. The Metal Trades Employers Association is found suddenly beginning to try for a more detailed central regulation. The background of this must be sought in the development of wages and manpower. For the employers' association the wage drift appeared as a threat to industrial profitability and organizational unity. The demand for control of development at the national level persuaded the association to accept a central regulation of all kinds of pay.

The roles of the employers and the workers, their endeavours and aims can be foretold with some degree of certainty proceeding from general knowledge. It is just as certain that the workers struggle to increase or at least to safeguard their earnings as it is that the employers tried to keep wage costs so los as possible. Thus comparatively speaking the parties' roles in the negotiation game are not complicated.

However, there were still facets of the rationalization question that led to further confrontations between the two parties. The policies of the organizations were created under the pressure of a powerful opposition, as in the case of the Metal Workers Union, or in contrast to other closely allied organizations in the case of the Metal Trades Employers Association.

The communist opposition against the leadership of the union was at times very powerful. In combination with the unaffiliated trade unionist aim, the opposition obtained a majority in representative organs for a certain period. In debate with this opposition the union's leadership was sometimes compelled to quite a far-reaching defence of rationalization and time and motion studies. This created chances for outside parties to establish definite proposals aimed at compromise. This possibility was used both by the conciliation commission and the Labour Court.

The Metal Trades Employers Association's internal settlement did not bring the organization closer to the workers but rather the opposite. The association was out to normalize the view of the problems of time and motion studies in the metal trades industry. This ambition was transferred to those who as the pioneers of the ideas of rationalization in Sweden wanted to present the time and motion study men, the engineers, as an impartial group serving both contestants and social development at one and the same time.

The part played by the Labour Court deserves to be noted specially. Through the court's attitudes conditions were created for the behaviour of the parties, which otherwise could only have been established with great difficulty. With a freedom greater than that of the contestants and with an authority greater than a conciliation commission, the court was able to produce constructive solutions of the problems surrounding the pay guarantee. At the same time the judgments throughout meant a powerful stimulation for the efforts towards rationalization and for a policy of compromise. Undoubtedly the part played by the Labour Court signified an increased professionalization of the relations between the two parties in the sense that the court's expert advice was requested; then demanded and presumed by the two parties. This had the effect that the lawyers, the interpreters, became a necessary medium for contacts between the parties. The role of the experts in this context developed parallel with another; that of the job study men.

Even in the case of job studies it is quite clear that the agreement at the end of the 1930s had become so complicated that it could not be grasped by the appointed representatives either at the lower or higher levels. Demands for expert help became increasingly loud. The time and motion study committee of the Metal Workers Union and the program for studies proposed by it is a sign of this, and a similar committee of the Metal Trades Employers Association and the increased attention it gave to the training of job study men is another.

The foundation on which the experts' roles was built was the same: the ideas of rationalization about objective, or intersubjective, correct, optimum part-solutions that only left small remainders for dispute between the parties. This foundation formed an important factor in enabling rational compromise at the central organizational levels to prevail over conflicts of interests whose concrete, tangible expression appeared only on the factory floor.

PART 3

The Textile Industry

CHAPTER 7

The textile industry—a presentation

1. Development of the textile industry

The Swedish textile industry can be divided into the following branches: the cotton, woollen, linen and hemp and jute, and the knitwear industries. Totally the textile industry was responsible for a good 6% of the whole of the sales marketing value of Swedish industry during the interwar years and somewhat less during the latter part of the 1940s. During these years the textile industry employed from 7 to 9% of those employed in industry and during the 1940s following the war somewhat over 6%.

As appears from the statistics presented below in Table 11, during the period shown the cotton and woollen industries were the completely dominating branches in the textile industry. The linen industry's share decreased steadily whilst the knitwear industry represented one of the interwar period's new branches with a very powerful expansion though perhaps somewhat slower during the first years of the 1930s.

Development of the production value in current prices, number of workers and motive power installed per employee appears from Diagrams 9 to 12. The reservations on the basic data and presentation made in connection with the presentation of the metal trades industry should also be made here (see pp. 27 et seq.).

The textile industry was essentially aimed at the domestic market and as one would expect does not reflect the slump of the 1930s to such a marked degree as did the metal trades industry. The decline in production of 1931 in the big branches was not proportional to the changes in the number of employees and this may largely be explained by the more than two-month conflict at the beginning of 1931. The knitwear industry with its steep growth curve appears most sensitive to competition; home sewing was probably more handy than home weaving in bad times. Hemp and jute were products used in other industries so that the branch was affected to a greater degree by the declining trade conditions.

Table 11. *The textile industry in figures*
Source: SOS Industry

	1925	1930	1935	1940	1945	1949
Sales mfg. value in million Skr						
Cotton industry	108	113	138	193	277	438
Linen, hemp, jute	23	22	21	35	41	61
Wool	107	105	119	195	230	402
Knitwear	35	61	73	126	161	256
Total	273	301	351	549	709	1 157
Number employed 1 000s						
Cotton industry	15	17	19	20	19	19
Linen, hemp, jute	4	4	3	3	3	3
Wool	13	13	14	16	14	16
Knitwear	5	9	11	13	12	12
Total	37	43	47	52	48	50
Production value per working hour Skr						
Cotton industry	4.14	3.90	4.43	6.17	9.34	16.23
Linen, hemp, jute	3.38	3.26	3.91	6.84	8.02	12.96
Wool	5.47	5.25	5.38	8.70	11.69	17.13
Knitwear	3.85	3.68	3.59	5.38	8.09	12.53
Motive power (h.p.) installed per employee						
Cotton industry	2.77	2.82	2.83	3.22	3.76	4.47
Linen, hemp, jute	1.93	1.89	1.75	1.90	2.55	2.81
Wool	2.21	2.37	2.32	2.54	3.26	3.56
Knitwear	0.56	0.45	0.45	0.49	0.72	0.86

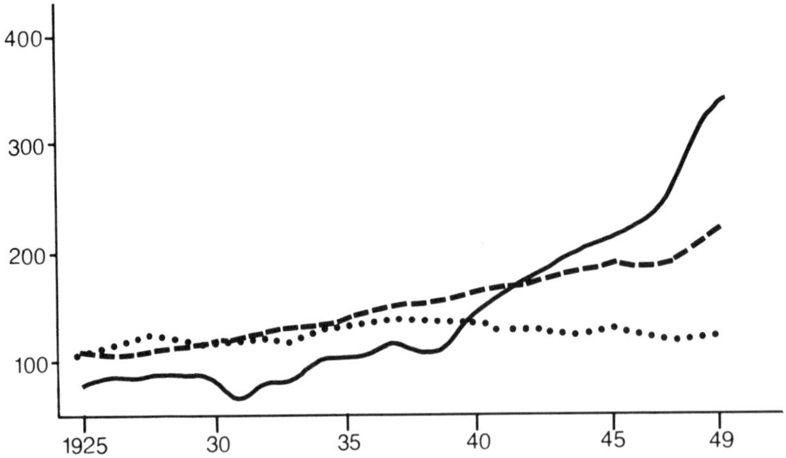

Diagram 9. Cotton industry.
Production value (—), No. of workers (· · ·) and power installed for direct production (---). Index: 1920–1924=100.
Source: SOS Industry.

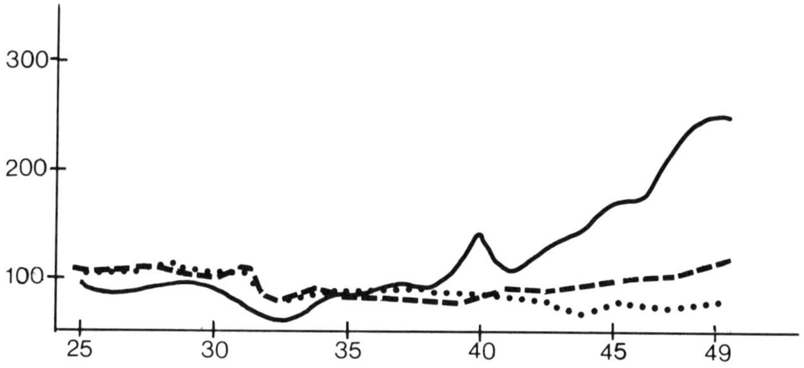

Diagram 10. Linen, jute and hemp industry.
Production value (—), No. of workers (· · ·) and power installed for direct production (---). Index 1920–1924=100.
Source: SOS Industry.

The curves on motive power installed indicate increasing mechanization during the period, especially in the knitwear industry where growth in production from the beginning of the 1940s is supported by a markedly increased utilization of machinery.

The development in productivity shown in Diagram 13 largely follows that of the metal trades. In the cotton and woolen industries the 1940s especially were a period of greatly increasing productivity even though here the development of prices tends to distort the impression.

The relationship between human and mechanical contributions is illumi-

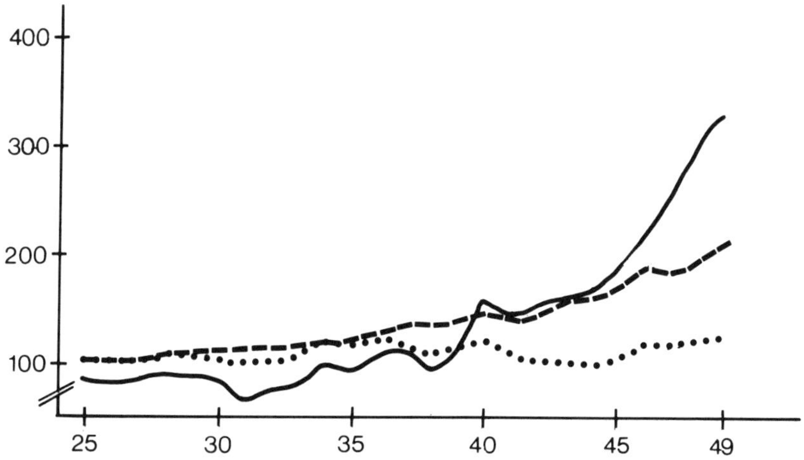

Diagram 11. Woollen industry.
Production value (—), Number of workers (· · ·) and power installed for direct production (---). Index 1920–1924=100.
Souce: SOS Industry.

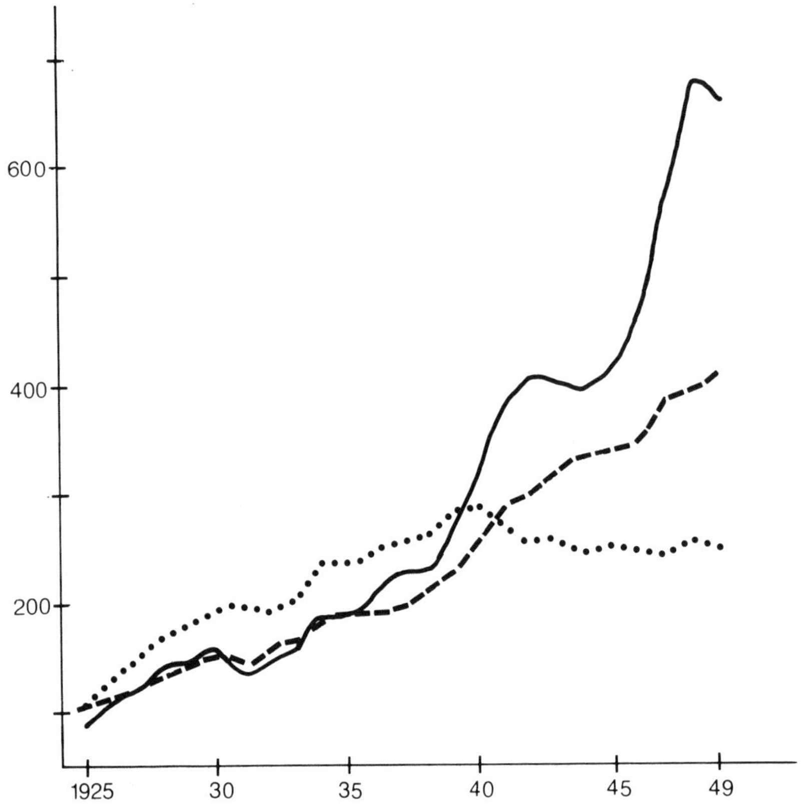

Diagram 12. Knitwear industry.
Production value (—), No. of workers (· · ·) and power installed for direct production (---). Index 1920–1924=100.
Source: SOS Industry.

nated by Diagram 14. In cotton manufacturing mechanization was pushed somewhat harder so that more horse-power was installed per worker but otherwise the relation was rather constant. The difference in level between the knitwear industry and the other branches probably more illustrates the differences in the need for heavier machinery than in the relation between men and machine.

The reversed difference exists with regard to another criterion of the degree of rationalization, the proportion of administrative personnel in the total number employed (Diagram 15). This diagram shows how conditions for the vertical division of labour increased within all branches of the textile industry during the whole of the period concerned. The knitwear industry appears most modern, the most intelligence-intensive or most specialized. However, in these respects the whole of the textile industry lies clearly below the average value for the whole of industry.

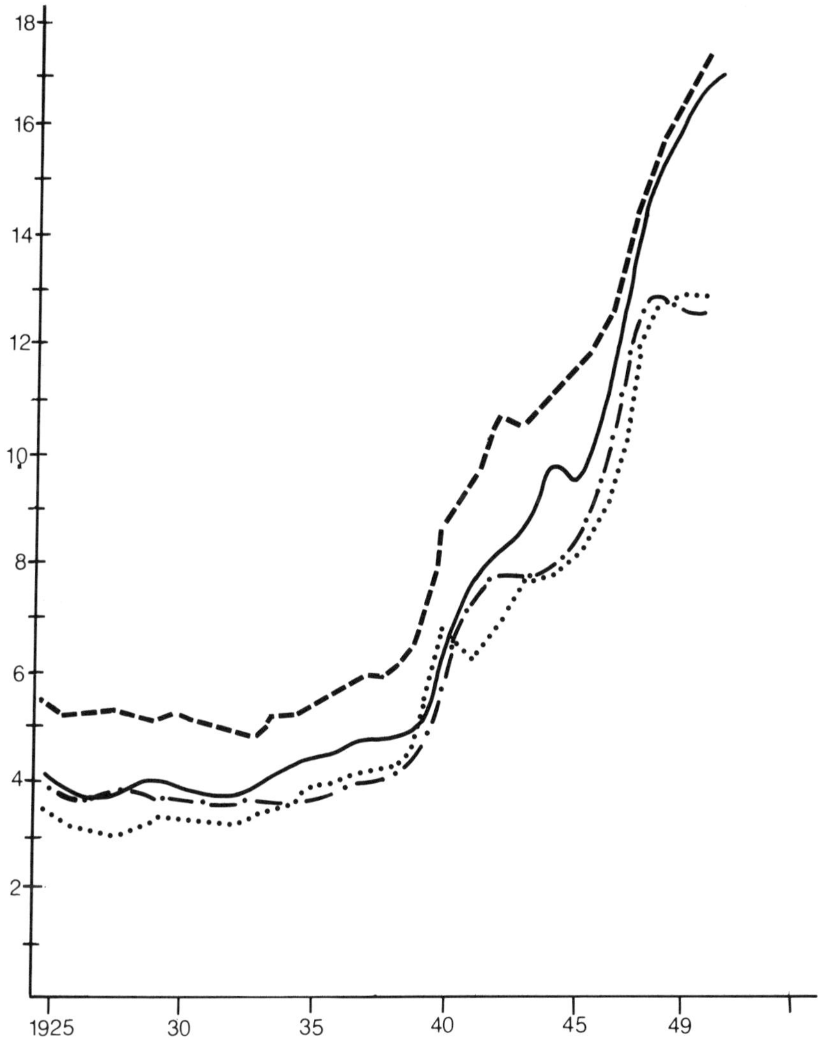

Diagram 13. Productivity in the textile industry.
Knitwear (–·–), Linen, jute, hemp (· · ·), Cotton (—), Wool (---). Production value per working hour. Skr. *Source:* SOS Industry.

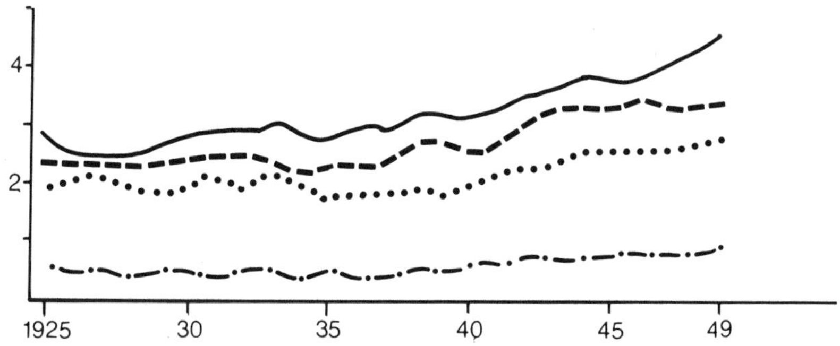

Diagram 14. Mechanization in the textile industry.
Knitwear (–·–), Linen, jute, hemp (· · ·), Wool (---), Cotton (—). Horsepower installed for direct production per employee. *Source:* SOS Industry.

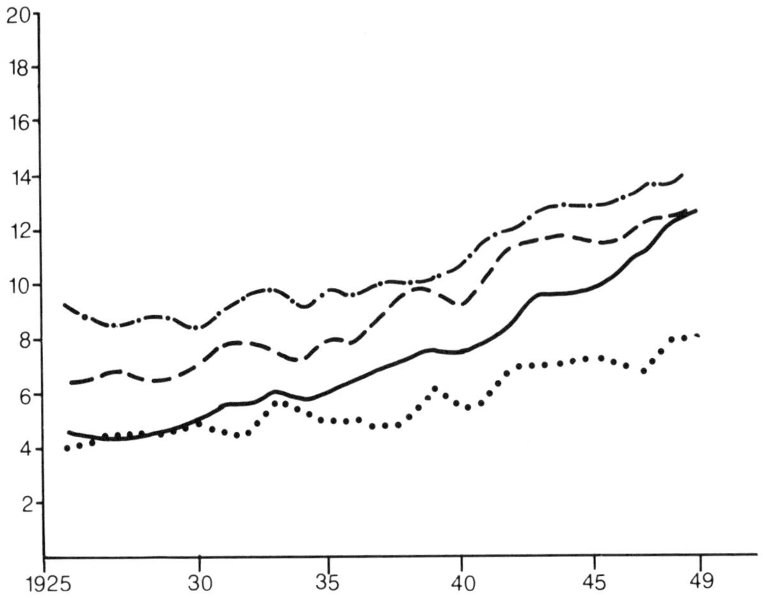

Diagram 15. Textile industry.
Proportion of administrative personnel as a percentage of all employees. Linen, jute and hemp (· · ·), Cotton (—), Wool (---), Knitwear (—·—).
Source: SOS Industry.

2. Pay and time and motion studies in the textile industry

Working conditions in the textile industry were regulated during the greater part of the period investigated by three different agreements: one that only applied to the textile workers, one for some specialist workers and one for tacklers. Only the actual textile workers' agreement has been investigated here. The pay for the true textile workers could be paid either as hourly pay or as piece wages. As the work on piece wages in accordance with agreement already gave higher earnings there was an effort by the Textile Workers Union during the whole of the period investigated to maintain and increase the proportion of piece wage hours in the total hours worked. Some figures from the literature are presented in Table 12 to show the size of the so-called piece wage volume.

Time wages were stated in the agreement as hourly wages expressed in öre in schedules for five different cost-of-living regions. Schedule 3 corresponded approximately to the average for all schedules.

The nominal hourly pay according to the agreement in schedule 3 during the period appears in Table 13. A real value of the nominal amount calcu-

Table 12. *Piece wage volume in the textile industry. The values apply to adult workers*

Source: Bergsten 1957 p. 363

Year	Male	Female	All
1938	42%	75%	55%
1950	66%	78%	71%

lated on the basis of the cost-of-living index is also included. The average hourly earnings are also given for comparison, that is to say an average of all wages including piece wage earnings, with both nominal and real values.

From Table 13 certain tendencies appear quite clearly in the development of the textile workers' wage conditions. The male hourly wages were practically unchanged during the first half of the period to then rise nominally by approximately 120% between 1937 and 1950. The real wages developed during the first half of the period somewhat more positively but the more obvious rise came at the middle of the 1930s and the mid-1940s.

The picture of the development of the average hourly earnings shows that the piece wage workers had a more favourable earnings' development than the hourly wage workers. Over the whole period the former managed to increase their real earnings considerably so that even the combined figures, which the average earnings constitute, show a doubling. The gap between workers on hourly wages and piece wage workers increased especially during the 1940s.

The comparison between wages for men and women shows the position of women as a low wage group, but also that the women had a definitely better real-wage increase, especially those who were on hourly pay. The nominal gap between hourly figures for men and women actually decreased during the second half of the period. Up to 1936 women's wages were about 65% of the men's; in 1940 they had risen to 75% and in 1950 to 80% of the men's. The female piece wage workers showed an even better earnings development although the difference between nominal wages and average hourly earnings for the female workers was not so great as for the men. The explanation of this may lie in the fact that the women's piece wage volume, already high at the beginning of the period, did not achieve the same growth as the men's.

That job studies were introduced into the textile industry in Sweden during the 1920s and intensified during the following decade to become during the 1940s an almost obvious part of activities is quite clear.[1] However, it is not possible to discover any real measurement of how extensive the use of job studies was at various times and at plants of various sizes in different branches.

For this purpose an attempt has been made to utilize the primary material

Table 13. *Wage conditions in the textile industry 1921–1950*

Source: Bergsten 1957 pp. 349, 352 and 353

Hourly wages according to schedule 3 including the appropriate high cost-of-living bonus, piece wage compensation and adjustable bonuses. All figures refer to adult workers. The average hourly earnings refer to earnings regarless of the kind of pay. Nominal amounts are expressed in öre (100 öre = 1 Skr), the real values in the index figure, where 1914 = 100.

	Men				Women			
	Hourly wages		Average hourly earnings		Hourly wages		Average hourly earnings	
Year	Nominal	Real	Nominal	Real	Nominal	Real	Nominal	Real
1921	110	134	114	137	71	147	74	132
1922	80	121	85	126	52	133	61	132
1923	77	128	86	140	50	141	62	149
1924	77	130	86	143	50	144	62	151
1925	77	128	86	140	50	142	62	149
1926	77	131	86	144	50	145	64	159
1927	77	132	87	147	50	146	62	155
1928	77	132	87	148	50	146	63	157
1929	77	134	88	150	50	148	64	160
1930	77	138	89	157	50	152	64	166
1931	77	142	90	164	50	157	63	170
1932	77	145	90	167	50	150	64	175
1933	77	148	90	170	50	163	64	178
1934	77	147	90	169	50	162	64	177
1935	77	145	92	171	50	160	65	178
1936	77	143	93	169	50	158	66	179
1937	88	160	99	177	59	182	70	182
1938	88	156	100	175	63	178	71	184
1939	92	158	106	178	70	184	75	187
1940	100	153	118	177	75	182	84	187
1941	107	144	127	168	83	172	92	180
1942	118	147	140	172	87	176	101	183
1943	121	149	146	176	92	182	106	189
1944	126	154	153	184	95	191	112	199
1945	134	163	164	197	99	205	121	214
1946	144	174	181	215	127	224	134	236
1947	162	185	211	237	129	247	160	267
1948	170	192	230	254	136	261	177	290
1949	170	189	237	258	136	257	183	295
1950	170	186	243	261	136	253	189	300

of the Rationalization Committee's questionnaire, the so-called representativity investigation.[2] This committee was appointed by the government to investigate the need for sociopolitical measures caused by rationalization industry. In 1936 the committee sent 348 industrial companies a questionnaire that also included a question of whether time studies or work studies had been used in the plant.

Of the questionnaires answered, 32 of the companies were in the textile industry. Of this 32, half replied definitely that time and motion studies had not been used. The question was unanswered by 11 companies. In most cases this can certainly be understood to mean there was nothing to report; time and motion studies had not been used. But in two of the cases there is certainly another reason. This applies to the answers from two large companies, Borås Wäfveri and Sveriges Förenade Trikåfabriker. These companies were also included in the committee's intensive investigation, as it was called, and supplied various information in this connection. Through this and other channels it is known that job studies were utilized at least from the beginning of the 1930s in both companies. Five companies including the largest in the woollen industry answered the question in the affirmative; time and motion studies were used in the company.

To sum up this brittle investigation material provides the following picture of time and motion studies in the textile industry at the middle of the 1930s: time or work studies had been introduced into all branches of the textile industry but were utilized preferably in the larger companies. It is thus obvious that the larger factories were also leaders in that section of the technical field dealing with the introduction of time and motion studies.[3] It is therefore of interest here to refer briefly to the picture how job study activities were introduced at Borås Wäfveri, which the job study manager of the company, Gunnar Koch, presented at a jubilee of the company in 1945.[4] The account looks just as though it had been cut from a Taylorian public relations publication.

At the beginning the survey centred on the person of Alex Engblom. He was like a pioneer who brought the light from the USA, the Taylorian teachings, but it was met by a compact lack of understanding in the Swedish environment. Time studies could not be used in the textile industry said, for example, Patrik Rydbeck at SKF, another of pathfinder companies for job studies in Sweden. The management at Borås Wäfveri where Engblom was employed at the beginning of the 1920s was, however, forward looking and positively inclined.

During the 1920s under the management of Engblom and the control of Axel Bergengren, a thoroughgoing technical rationalization was carried out in the company, in which time studies and method studies were used. In 1931 a special central time and motion study department was set up under the management of a time and motion study engineer from SKF, Gothenburg. Up to the middle of the 1930s a systematic examination was carried out of piece wages in the spinning and weaving departments of the company. In these measurement and analysis of yarn breakage were also used, which had not been customary earlier.

At the beginning of 1936 Gunnar Koch was appointed manager of the time and motion study department. Koch had already been employed earlier in the company and during 1935 had been a student at the first course for time and motion study engineers held by the Institute for the Training of Foremen.

During the latter part of the 1930s time and motion studies were concentrated on putting the work at the preparation plant on piece wage basis. Many attempts were made to offer piece wages on the basis of time and motion studies but they were met with suspicion by the workers and were unsuccessful. This had the result, however, that certain jobs were transferred to female labour—a saving. In the Taylorian spirit a splitting wedge was then inserted, a small operation at the printing shop, "the sewing of primary material".

All seamstresses were transferred to one department where it could be seen at once that the number of workers was too large. The machine park was renewed, the workplaces were arranged in the best manner and first class material was provided at a satisfactory speed. Following this the work was time studied and a piece wage was paid as a test. This showed that it fulfilled its promise with regard to earnings and was confirmed. When other workers in the printing shop saw how well the working conditions had been arranged for the seamstresses they also wanted to have their work studied and before the end of 1939 all departments within the company had been investigated with time and motion studies.

Many charactheristics of the development described by Koch are worth noting. There is time setting: time studies had been utilized essentially as method studies during the 1920s but from the beginning of the 1930s, presumably in connection with the revision of piece wages in 1931, formed the basis for the piece wage offer. It was the communication of ideas and the educational aspects that confirm the picture I have presented in another context: the role of the pioneer with ideas from the USA, the foothold in the metal trades industry that brought the first time and motion study engineers, a role later played by the Institute for the Training of Foremen (*Arbetsledarinstitutet*).[5] There are to be found at the beginning the workers' suspicious reactions to the results of time studies: the replacement of male labour by female labour, a decrease in the number of seamstresses in conduction with centralization. There are also changes in the workers' ideas when certain groups of workers could show improvements in pay and working conditions.

Transfer of the practice of job studies from the metal trades industry to the textile industry was certainly not simple. In a report of 1935 about work studies in the textile industry Alex Engblom devoted a great deal of space precisely to these problems.[6] A decisive difference between the textile and the metal trades industries in this particular respect was, thought Engblom, associated with the material processed. In the metal trades industry metals were dealt with whose characteristics were well-defined and whose working could therefore be carried out in accordance with a very regular schedule. In the textile industry on the other hand a living raw material was involved, a fibre, whose characteristics certainly could be stated as an average value on a scale (for example length of the fibre) but which showed a very great spread around that value. Irregularities in the material created considerable disturbances in machine operation. Often more than half of working time was used to clear such stoppages, such as those caused by yarn breakage. Irregularities in the material and irregularities in breakdowns had the effect that the work factors considered to form part of an operation could certainly be distinguished, but their frequency was difficult to foretell unless an extensive time study was made. These circumstances also led to a concentration in the textile industry on lost time studies in the shape of yarn-breakage analysis in order to decide the characteristics of the material—its degree of wearability, which was included as an important factor in the different formulae constructed for the setting of piece wages. The demand for long-term studies and the comparatively modular form of working operations in the textile industry actually had the effect that the number of piece wages subjected to job studies was considerably lower than in the metal trades industry. Nor does there appear to have been any performance index used, implying a discretionary evaluation of the studied workers' industry and skill. One reason for this was the comparatively greater dependence on the characteristics of the material and the provisions of the existing agreement concerning production investigations. It was laid down in this agreement that normal conditions should exist and that the production of workers of average skill should be studied. This was later also to apply to time studies.

3. The two sides

The Swedish Textile Employers Association (Sveriges Textilindustriförbund) was formed and obtained admission to the Swedish Employers Confederation in 1907.[7] The association's largest organ for decision-making was the general assembly which met annually in order to examine the administration of the general council, carry out certain elections and decide on the association's levy for membership fees.

The association was organized in comparatively independent districts that gradually, especially after the principle of national agreements had come into being, transferred increasing powers to the common general council. This consisted of 10–12 members so that the different districts as well as the various branches were represented. In the general council a working committee was set up consisting of the chairman, vice chairman and the managing director.

Table 14. *Number of member companies (a) and number of workers employed (b) within the Textile Employers Association*

Source: Bergsten 1957 p. 159

	1909		1928		1950	
District	a	b	a	b	a	b
Southern district	21	2 859	17	3 457	43	9 331
Borås district	28	3 718	31	8 890	43	10 697
Gothenburg district	29	7 337	30	9 076	59	13 140
Norrköping district	31	5 538	23	5 600	46	7 043
Northern district	19	3 248	20	4 095	52	6 545
Whole association	128	22 700	121	31 136	243	46 756

For practically the whole of the period covered by the investigation the director of Borås Wäfveri, Axel Bergengren, was chairman of the Textile Employers Association (1926–1945). He was followed by Nils Engberg, a director at Skånska Yllefabriken in Kristianstad.

The association's office was managed from the beginning by an ombudsman. This post was reconstructed in 1920 to that of managing director. From 1907 up to 1944 this post was filled by Erik Paues. He thus came, together with Bergengren, to dominate the employers' actions during the period with which we are concerned here. He also assisted in the setting up of the Swedish Association of Clothing Manufacturers (Sveriges Konfektionsindustriförbund) in 1920, where he became both chairman an managing director.

Paues was succeeded—in practice as early as 1939—by Otto Thurdin, a lawyer like his predecessor, and since 1930 secretary of the association. Thurdin led the activities of the Textile Employers Association up to his death in 1948.

The number of members in the association for certain years is shown in Table 14.

The Swedish Textile Workers Union (Svenska Textilarbetareförbundet) was established in 1898 and admitted to the Swedish Trade Union Confederation two years later.[8] The first decade of the organization was marked by a varying degree of success and for various reasons, including financial ones, the union resigned from the Trade Union Confederation in 1912 and did not rejoin until after the great conflict in 1931.

Between 1906 and the end of 1930 Gustaf Janzén headed the union and became a legendary dominating figure. He was followed during 1930–1933 by K. J. Olsson who later, via the post of chief editor for the newspapers Västgötademokraten and Ny Tid, finished as the county governor of Örebro. Olsson was followed as the appointed representative by Gösta

Table 15. *The Swedish Textile Workers Union. Number of members and number of branches*

Source: Tillman 1948, p. 122 et seq.

Year	Members	Branches	Year	Members	Branches
1920	17 577	96	1934	32 380	119
1921	15 471	92	1935	32 419	117
1922	15 736	96	1936	33 998	110
1923	17 304	99	1937	34 971	93
1924	20 043	105	1938	34 931	90
1925	21 574	110	1939	36 954	94
1926	22 634	108	1940	38 658	99
1927	23 966	108	1941	39 222	102
1928	24 857	110	1942	39 668	101
1929	25 273	112	1943	39 413	102
1930	27 185	121	1944	40 175	103
1931	30 085	122	1945	41 534	102
1932	30 130	120	1946	42 154	102
1933	30 798	122	1947	43 035	106

Wennström (called chairman after 1936), a tackler from Norrköping, later to become during the 1950s a commissioner of Stockholm.

The Textile Workers Union held its Congress every fourth year. This was the union's highest organ for decisions and consisted of representatives from the branches elected directly by the members. Between congresses the representative council, after 1936 renamed the general council, met biannually and this also had to decide whether notice should be given to terminate the current agreement. Current matters were dealt with by the executive board, which was increased to 9 members in 1936.

Development of membership in the union is shown in Table 15.

In addition to the Textile Workers Union there were also a number of other unions, which also concluded agreements with the Textile Employers Assn. They were either special unions or larger unions that also organized some smaller category of workers within the textile industry: The Amalgamated Unions (De Förenade Förbunden), the Swedish Federation of General and Factory Workers (Grov- och fabriksarbetareförbundet), the Swedish Machinists, Stokers and Repairers Union (Maskinist-, Eldare- och Reparatörsförbundet, later the Driftspersonalsförbundet) and the Swedish Loom Tacklers Union (Vävlagarförbundet).

4. Development of the textile agreement

The first national agreement for the textile industry was concluded between the Textile Employers Association and the Swedish Textile Workers Union on 1 June 1914. It was valid up to the end of 1916 when a new agreement was made. A second agreement was concluded in January 1920, this time for a year but prolonged with regard to the general provisions for a further year. In February 1922 a new agreement was signed that was in force until the end of 1924. The 1925 agreement then applied up to the end of 1930. The collective negotiations preceeding the next agreement were to be the hardest in this field with an industrial conflict that lasted two and a half months at the beginning of 1931. The general provisions of the 1931 agreement applied up to and including 1936. In 1937 a new 2-year agreement was made, which for most of the general provisions was prolonged and these apply to the end of the period investigated. Smaller changes with regard to the piece wage paragraph, however, were carried out in 1939 and 1947.

The provisions of the textile agreement concerning piece wages are to be found for the whole of the period covered here in paragraph 7 of the agreement. As Bergsten points out, the size of this paragraph increased from about half a page in the first agreement to about three pages by the middle of the 1950s, an indication of the degree to which the individual piece wage negotiations had been included during the adjustment of the national agreement![9]

The first national agreement, signed in 1914, counted with two kinds of piece wages, those based on piece wage schedules (regulated in clause 1 of the paragraph) and the remainder (clause 3). Piece wage schedules had the same period of validity as the whole agreement. Thus it followed that piece wages included in such schedules could not be altered unilaterally by the employer. So that the system should not be too rigid a "rationalization provision" was included in the piece wage paragraph, clause 2. If changes in machines, work routines or working material were made that changed the conditions for the calculation of piece wages, then the piece-wage price could be adjusted. The fixed piece wage schedules should thus not inhibit development within the industry.

In order to guarantee the workers against misuse of this possibility of piece wage adjustment, it was of course carried out unilaterally by the management, an earnings guarantee was tacked on the section. A new piece wage should not be set lower than that at which the worker with unchanged skill and labour input should be able to maintain his earnings.

For other piece wages not included in the piece wage schedule there was no comparative guarantee for the earnings level. The piece wage should be determined by free negotiations between management and the worker or workers who were to work on it. If the worker or workers found that the piece wage offered was not such as they could accept then the work should be carried out on hourly pay.

At that time, when the period of the present investigation begins, the arrangement of the paragraph in the agreement was the same, but in influence on the setting of piece wages the workers had advanced their positions. The right to transfer disputes about piece wages to central negotiations was laid down in the agreement of 1920. Thus the organizations had now the chance of influencing the setting of piece wages when a dispute arose. This was also to apply in cases where the employer wished to alter piece wages listed in a piece wage schedule. Reference to changed conditions was no longer to be sufficient for the management alone to alter fixed piece wages. As the organizations had thus been assured of influence the earnings guarantee, as laid down in the agreement of 1914, was removed as a quid pro quo.

The earnings guarantee was soon to be reinstated in the agreement's provisions by means of the regulation of the piece wage excess. Rules for the calculation of piece wages were included in the agreement of 1922. The excess of piece wages over hourly pay was stated to be *on average 15%*. The hourly pay for the highest wage class should be used as the basis of calculation. In addition there were provisions on how to proceed in the setting of piece wages for work carried out by both men and women. Such piece wages should be set on the basis of the women's hourly pay, but the men should be given the possiblity of increasing their earnings in some way.

By means of an adjustment in the text of the agreement of 1925 it is made still clearer that the basis of calculation should be applied not only to the fixed piece wages but also to the others.

Rules about at least a 15% piece wage excess remained unchanged in the agreement up to 1947, when the piece wage targets *in öre* replaced those stated as a percentage. These were calculated from the 15% excess but the change implied standardization across the area of the agreement. The textile agreement had thus obtained cash factors laid down by agreement for the calculation of piece wages.

In the setting of piece wages a difference was made between the piece wage entered in the piece wage schedules and the others. The schedule was in principle given the same status as the agreement: it had the same period of validity and was confirmed by the organizations in connection with the signing of the agreement. The other piece wages were left for free negotiations at the local level and influence from the organizations could first be brought into play when a dispute arose. This division into piece wage categories was to continue in principle.

In the agreement of 1931 the status of piece wages was emphasized still more strongly. The piece wage schedules were to be approved by the main organizations and all alterations in the schedules motiviated by rationalization measures should be reported by or approved by them. A further step in this direction was made in the agreement of 1937 where it was laid down that a worker, who in piece wage negotiations considered that the piece wage should be added to the fixed schedule, had the right to consult his union. This meant that the local union representatives had the right to help right from the start in individual negotiations to that degree these concerned additions to the fixed schedule.

However, the division into categories of fixed and other piece wages was not so simple. Perhaps the growing wall of provisions around fixed piece wages also compelled other kinds of local agreements.

The text of the 1931 agreement had included a possibility of local agreement on piece wage schedules with a shorter period of validity than that in the national agreement. Such agreements would not need to be approved by the main organizations.

In the text of the agreement for 1937 the difference became explicit. Then were found again, in paragraph 7, three different categories of piece wage agreements. There were the fixed piece wage schedules, which applied for the same period as the national agreement and were approved by the main organizations and also the other piece wages, oral agreements, which only applied until further notice and were subject to a week's notice. In addition there was also a new category, the provisional piece wage schedule, which had a shorter period of validity than the fixed ones and did not need to be submitted to the central organizations for approval. The difference between the provisional piece wage schedule and the other piece wages was that in the former case the local union organization participated in the piece wage agreement; but in the latter the management made an agreement directly and solely with the workers concerned. During the remainder of the investigation period the three different kinds of piece wages were always to be found in the text of the agreement. The piece wages agreed verbally, however, constituted a very small part of the piece wage volume and after 1937 the development gradually moved away also from the fixed schedule. Most of them became regulated in accordance with provisional piece wage schedules.

In the above the provisions intended to guarantee the earnings level of piece wages have been dealt with and also those intended to regulate the degree of influence over the setting of piece wages that management, the workers and their respective local and central organizations were to have. These provisions and the changes shown in them may of themselves be said to be independent of how the piece wage was calculated. From 1931 provisions were also introduced that regulated the basis of piece wage setting. It was then laid down that investigation of the production that formed the basis of piece wage setting should be carried out if either side so requested. Thus even the workers had the right to take the initiative. It should also be observed that the local union organization was ensured the right to take part in or check the investigation.

The meaning of the term 'production investigation', as appears from the following investigation, was to be interpreted differently by the two parties. The Textile Workers Union, which took the initiative for the provision, meant a statistical, simple calculation on the basis of production actually carried out and thought that the provision in the agreement excluded the use of, for example, time and motion studies. The employers did not accept this interpretation and at the end of the period covered by the investigation it is obvious that both parties regarded the provision on production investigation as a valuable regulation of just the conditions of time and motion studies.

A time limit of a month was laid down for the carrying out of a production study. But even with such a limitation the position was of course that the workers' earnings during the waiting period for a new piece wage had to be regulated. In the agreement of 1937 it was prescribed that during such an interim period the worker should be guaranteed a bonus above the hourly pay of 10%. This compensation was included in the following agreement of 1939 up to the minimum level of the piece wage excess, 15% above the hourly pay.

A similar situation ensued if a dispute arose because the workers were not prepared to accept the piece wage offered by the employers. In the first national agreement in 1914 it was laid down that in such cases the work should be carried out at hourly pay. In the agreement of 1920 and onwards this provision stood fast but, however, only after a round of negotiations between the central organizations. In the agreement of 1937 there was also the possibility that the work, following dispute and central negotiations, was carried out *either* at hourly pay *or* at the piece wage offered by the employer (which naturally in any case must be considered to lie at least 15% above the hourly pay).

To sum up:
The first national agreement in the textile area was concluded in 1914. The most important changes in the piece wage paragraph were carried out after this in the agreements of 1920, 1922, 1925, 1931, 1937, 1939 and 1947.

Types of piece wages:
The provisions of the agreements dealt with different kinds of piece wages. Already in the first national agreement a difference was made between those listed in piece wage schedules and other kinds. Indicated in 1931 and expressly in the 1937 agreement, a difference was made between fixed and provisional piece wage schedules.

Influence:
The influence of the trade union organizations expanded during the period covered by the present investigation and was also dependent on the kind of piece wages. In disputes about all kinds of piece wages the organizations could be called in. All piece wages entered in the fixed piece wage schedules (those with a period of validity which agreed with that of the current national agreement) were to be approved by the main organizations, which should also be informed if notice was given to terminate such piece wages. From 1937 the local unions received the right to participate as early as at the initial piece wage negotiations when these dealt with such agreements as were intended to be entered on the piece wage schedule.

11−812400 De Geer

Earnings-level:
With regard to changes in the piece wage schedules, one form of guarantee of the earnings level was already to be found in the 1914 agreement. From 1922 the piece wage excess on all piece wages was stated to be on average 15%, which in 1931 was changed to a minimum of 15%. From 1947 the piece wage was expressed in öre. From 1937 the piece wage compensation was regulated, whilst the setting of the piece wage was awaited, to 10%. This was 15% from 1939.

Production investigation/time and motion studies:
From 1931 the grounds for the setting of piece wages were regulated in the agreement. The right to take the initiative for production investigation lay with both parties. The parties should then jointly select the workers to be studied and the investigation should apply to normal production and workers of average skill. The local trade union organizations had the right to participate and to check the studies. In spite of opposition from the workers time and motion studies came to be accepted in the textile industry and the provisions on production investigation to be interpreted as a regulation of these.

CHAPTER 8

The great textile strike

1. The workers' offensive

The national agreement signed under threat of a lockout in 1924 was prolonged during the years following and remained in force until 1930. As the 1924 agreement had meant the status quo for both hourly wages as well as piece wages, the wages regulated in the agreement were maintained at an unchanged level from 1922 to the beginning of the 1930s.[1]

The collective negotiations preceding the agreement of 1931 were started as early as March 1930. The representative council of the Textile Workers Union then decided to convene a policy conference in the autumn to discuss the agreement so that the branch representatives could decide whether the agreement should be terminated.[2] But before this the matter was taken up at a new meeting of the representative council.[3] Against the opinion of a strong minority (six votes to nine) it was agreed to convene the policy conference to discuss the agreement. The minority wanted the question to be balloted the among the members.[4] Soon afterwards the executive board duly appointed a new representative, K. J. Olsson, together with David Ekman and Gösta Wennström who were members of the board, to formulate a proposal for a new agreement to be put before the conference.[5] Notice of the conference was sent out at the beginning of August and included a suggestion that delegates should not come with restricted mandates but after the discussions at the conference were free to decide on the question of notice.[6] The suggestion was accepted. There was almost a full muster and only 5 of the union's 117 branches were unrepresented.[7]

The policy conference was called for Saturday 27 September 1930. Two days earlier, following a discussion point by point, the executive board had approved the draft agreement which those appointed had presented.[8]

First in the line of speakers at the conference was K. J. Olsson. Methodically he went through the arguments for and against giving notice. He emphasized the importance of them being so 'objective' as possible in their assessments. He thought the reasons that could justify issuing notice were initially the obscurely written and out of date general provisions. On the other hand, with regard to wages it was more difficult for the textile workers to get increases even though the situation was difficult for the workers on hourly wages. The situation in the textile industry was not especially favourable because of difficult competition with countries outside Europe and the care exerted by Swedish manufacturers because of the falling prices of raw materials. In such a situation it was a case of not manufacturing for stock from a raw material the price of which was soon to be considerably lower. He then reported on unemployment in the textile industry and about working conditions otherwise. He thought the intensity of work had been increased and also that rationalization had caused a concentration of operations. The tone of Olsson's speech was very careful throughout. He avoided any clear personal stand, though a sensitive ear could not mistake his meaning as it appears in the account of his speech:

> He could well understand that among many members of the union the idea existed that the union should take measures to improve working conditions if possible. Yes, it could even be thought that there were members who had the idea that the risk of a reverse should be taken rather than that nothing should be done. However, the position is ... not especially favourable ... With regard to the chances of being able to obtain any improvements by giving notice to terminate the agreement, they were not particularly great. Perhaps there was a chance of an improvement in the general provisions and possibly even some improvement of the hourly paid workers' position, but at the same time there also existed the risk that piece wages, which in many cases lay relatively high, could become the subject of cuts. In addition one could of course even calculate with a conflict, long or short, which obviously is of very great importance for the union ...[9]

Following Olsson's exposition about half of the representatives present availed themselves of the chance to speak. A clear majority requested that notice should be given. Only three speakers opposed termination and scarcely ten had no definite demands. Various grievances were advanced as reasons for termination. These referred both to the regional and the general provisions of the agreement. That rationalization had increased the work tempo was found among the arguments but the time and motion study problem did not appear at all in this context.

An acrimonious attack was aimed at the leadership by a representative from Stockholm who accused both the board and the representative council of being Mondists, from whom the workers could not expect anything goood. A radical attitude also marked demands about the agreement, which were advanced among others by a radical communist, John Berglund, on behalf of the Karlstad branch.[10]

The policy conference closed with a vote on whether the agreement should be terminated. The ballot papers were handed over for counting at the representative council's meeting the next day when it was found that the majority for giving notice was large, but not so overwhelming as the debate had suggested. A total of 81 representatives had voted for notice to be given and 30 against. One ballot paper was blank. Such a powerful opinion could not be ignored. Therefore the representative council decided that notice should be given to terminate the national agreement. Only one of its 17 members went against this decision.

Following this the representative council proceeded to examine the executive board's draft for a new agreement. On the points with which we are concerned in this context—paragraphs 7 and 15, see below—the draft was approved after a few remarks.[11]

The Textile Workers Union's proposal for a new wording of the piece wage paragraph of the agreement thus came to contain the following differences from the current text:

1. The provision of the "rationalization clause", that agreements on piece wages included in fixed schedules could cease to apply if the conditions of production were changed, was made more expressly *reciprocal*. *Either* party should in such cases have the right to terminate the piece wage.

2. In the new wording there were more compulsive demands that new piece wage agreements should be reached in order to replace fixed piece wages that had been terminated or to complete the price schedule if new goods had come forward. The piece wage volume should not be allowed to decrease.

3. With regard to the other (free) piece wages the agreement laid down that agreements should be made between the management and the workers concerned. It was now proposed that the agreements should be made between management and the trade union.

4. Women who were given men's piece wage work should also be paid according to the men's wages.

In addition to changes in the piece wage paragraph it is interesting in the context of this investigation to observe the innovations put forward in that paragraph of the agreement regulating the negotiation procedures, paragraph 15. This included the first expression of a reaction because time and motion studies had started to be used with the textile industry and this may be regarded as an attempt to regulate, apportion, control over the new technology. The text in the draft agreement read as follows:

4. If an investigation is requested from either side concerning the production that shall form the basis of the piece wage setting, such investigation shall be carried out. The investigation shall concern the average production of an average worker. The parties shall jointly select the workers whose production shall be investigated and the workers' party shall be given the opportunity of taking part in or checking the investigation.

The investigation may not continue longer than is required to carry out a satisfactory investigation and in any case may not proceed longer than one month.

What, especially in the light of the experiences revealed by the metal trades agreement, appears as remarkable in the proposed wording is how the trade union organization and management are put on an equal footing. Both parties shall have the right to take the initiative for production investigation, in joint consultation they shall select the workers on whom the production investigation shall be carried out, and both parties shall have the right to participate in or check the investigation. With such a provision the trade union organization would at once obtain a firm grasp on the basis of piece wage setting, a grasp that the workers side in the metal trades area had to fight for a little at a time.

But whilst the Metal Workers Union during the 30s and 40s gradually tried to gain influence over time and motion studies, whose justification was not brought into question, the Textile Workers Union's proposal about provisions concerning production investigation should be seen in another way. One important item in the proposed provision was thought to be that piece wages should be based on *actual production*, not on such *prognoses for future production* that the time and motion studies served to facilitate. The provision on production investigation was therefore not wholly a proposal to obtain control over how the basis for piece wages was obtained, but also a proposal that aimed at nipping things in the bud by means of provisions in the agreement to keep time and motion studies outside the textile industry.

Otherwise the workers' draft agreement contained an increase of wages by 25% plus an extra 15 öre per hour for hourly workers, a decrease in the number of schedules from 5 to 4, a 6-day holiday and an average piece wage excess of 30%.[12]

The proposed changes, the demands on which the Textile Workers Union went to battle, will here be condensed into different categories, in different lines for the workers' aspirations. There is a line for increase in wages, which was also expressed through a demand for a higher piece wage excess. There is also the endeavour to acheive a general increase of the piece wage volume and especially in the proportion of the fixed piece wage schedules. Also the ambition to increase the equality between the employer and the workers' organizations with regard to the rationalization clause and the basis of piece wages. Complicated systems, including time and motion studies, were to be excluded. There is finally an aspiration for equality between the sexes—as is known the textile workers had quite a large proportion of women workers in their ranks.

2. The employers' state of preparedness

On Monday 29 September the delegates of the Textile Workers' Union with their draft for the new agreement presented themselves to their opponents and delivered the written notice of termination of the agreement. The

Textile Employers' Association replied with an immediate counternotice.[13] Naturally the employers had not awaited completely passively the time allowed for giving notice. At the middle of September an extraordinary meeting had been held of all branches of the Textile Employers Association. At this meeting there had been complete unity that the employers should not take the offensive. At a conference of the representatives of the Swedish Employers Confederation held simulaneously it had been quite clear that neither did the other employer associations who might do so have any intention of giving notice to terminate their agreements. Because of this the general council of the Textile Employers Association decided to wait and see, to give notice only if the workers took the step first. To be prepared for this eventuality the council at the same time approved a proposal for a new agreement to be kept available.[14]

The employers' draft agreement included a demand for a 10% reduction in wages. There were also changes in the general provisions. The piece wage paragraph, number 8 in the draft, contained essentially the following alterations:

1. A provision that fixed piece wage schedules should not be contrary to the current national agreement. This appeared to be an obvious provision, but the idea behind it was to provide the association with the chance of preventing wage drift and competition for manpower by rendering void the piece wage schedules that gave a higher excess than those stated in the national agreement.[15]

2. Other agreements on piece wages made during the period of the agreement should unless otherwise stated have the same period of validity as the agreement. The agreements that thus received the same period of validity as the agreement should be added to the fixed piece wage schedules and be reported to the main organizations. This proposed alteration meant that piece wage schedules with a shorter period of validity could be introduced.

3. In disputes about a piece wage offered the work was to be carried out at hourly pay whilst the parties carried the dispute to further negotiations.

4. The rationalization clause received a softer wording. In the agreement then current clear conditions were stated: if new working methods were introduced, or if improved machinery was utilized, or if changes in the work or the product were present, then the piece wage agreed could be terminated. In the new proposal from the employers there was now to be found a much more sweeping formulation:

> If the conditions for a piece wage are altered with regard to working methods, machinery, material or otherwise, the party concerned has the right to give notice to the other party to terminate such piece wage, such termination to come into force one week from notice being given.

The proposed change thus meant a relaxation, an increased freedom of movement for the parties. It was obvious that this freedom could initially be utilized by the employer, who had the right to decide on how production should be planned. Even changes of another kind than those envisaged by the author of the agreement's text could by virtue of the wording of the provision lead to changes in the piece wage.

So that the freer wording did not mean losing a hold on its application in practice, the employers' proposal for a new rationalization clause continued with the provision that all notices of termination of the kind concerned should be reported without delay to the main organizations, who had of course already approved the piece wage agreements where notice had been given.

Lying in these two last named points were the most important new things in the employers' proposal. If they could be carried through important advantages would be gained. A piece wage dispute would not delay production and the work would be carried out at an hourly wage that in most cases would be cheaper. This also provided of course the possibility observed by the workers that the employers by offering "impossible" piece wages could compel the piece wage workers on to hourly pay and thus oppose the workers' struggle to increase the volume of piece wages. The more subtle wording of the rationalization clause would partly increase the possibilities of rationalization measures of all kinds and perhaps also increase the urge for rationalization as this offered a possibility of getting rid of the piece wages regarded as fixed.[16]

3. Initial negotiations

Both parties had thus given notice to terminate the current national agreement and it now was a case of utilizing the time before negotiations about the new agreement could be started.

On the part of the employers, the Textile Employers Association's general council decided at the middle of October to invite the Textile Workers Union as well as the other, smaller opposing parties (the Amalgamated Unions (*De Förenade Förbunden*), the Swedish Federation of General and Factory Workers (*Grov- och fabriksarbetare förbundet*), the Swedish Machinists, Stokers and Repairers Union (*Maskinist-, Eldare- och Reparatörsförbundet*) and the Swedish Loom Tacklers Union (*Vävlagareförbundet*) to negotiations in the first week of November. Then the whole of the council would take part in the negotiations.[17]

In a circular letter the Textile Workers Union reported to its members about the notice given and the demands put forward. This brought a reaction later on in a letter from the radical branch 36 in Karlstad, which expressed its satisfaction at the notice and pointed out that the demands lacked any scope for bargaining.[18] This, together with the invitation to negotiations in Stockholm was placed before the executive board of Textile

Workers Union, which selected a negotiating team.[19] The union, as the other organiations involved, accepted the employers' invitation.

During four days at the beginning of November the opening negotiations got under way, the chairman being Erik Paues.[20] The first days were spent in trench warfare, a general discussion about the situation of the industry and the conditions for the agreement. Erik Paues, for the employers, stressed the importance of the other side not only considering its own interests but also the prosperity of the whole industry. At the same time he demonstrated a typical current attitude to the problems dividing the two sides: if only the conditions for industrial and economic activities were sufficiently illuminated he was quite certain that the differences of opinion preceding the new agreement would be decreased considerably. The idea was thus that the different views held by the parties were not founded on definite conflicts of interest but were more the fruits of a lack of information and lack of insight into the conditions of the other side.

In his introductory speech the textile workers' representative had to give reasons for the notice of termination as the workers had taken the initiative. K. J. Olsson said the current agreement was out of date and this was sufficient cause for a thoroughgoing revision to be carried out. In addition the agreement was biased and took more notice of the employers' than the workers' interests. This of course reflected the balance of strength between the parties at a time when the general provisions had been made. In addition there were a number of provisions in the agreement that were so ambiguously written that a variable practice could stem from them. This was all the more dangerous, he said, as a different kind of mentality was beginning to spread among the employers. They had started to give more importance to interpreting the *wording* of the agreement with legal quibbling than to try to follow the *spirit* that existed when the provisions were made. The reason for this changed attitude was to be found in the passing of the Collective Agreements Act and the establishment of the Labour Court. These circumstances were the reason why it was now necessary to put a great deal more weight on expressing intentions in legally unassailable terms.

A further reason for notice being given was the delay which affected the textile workers' wages in comparison with other groups of workers. Hourly paid workers especially had been subjected to this. They had been compelled to take part in intensive rationalization and in the greatly increased speed of work without sharing in the profits that had arisen by receiving higher earnings.

Finally K. J. Olsson referred to a characteristic idea of the times, which was an echo from the national Mondist industrial conference two years earlier. It would be sensible of the employers, he said, to listen to the workers' demands in order to create greater job satisfaction among the workers. Compliance in the demands made for the agreement would be to the advantage of both parties.

The arguments put forward by K. J. Olsson were met by the employers' negotiating team with arguments that seem obvious in the context: certainly the mentality of the employers had not changed, but the new conditions of industry demanded greater precision; a textile worker's wages were certainly lower than those of other groups but this was the case everywhere in the world; real wages had increased although the nominal wages had remained stable for a number of years; in the families of textile workers it was usual that many members worked so that the family's total income was not so low; the approaching crisis called for carefulness. Instead of the increase requested the employers were therefore compelled to ask for a wage decrease.

The argument on the approaching crisis was rejected in its turn by the workers with reference to the fact that its effects had not been felt and that ideas about it were divided. Breadwinning for the textile workers was difficult. Nor for industry on the whole was it an advantage that wages were too low. For the higher the wages industry paid the greater the purchasing power created for the public.

Following this standard run-through of common arguments by which the opposite party was never convinced, during the following days the negotiation teams went through the general provisions in both draft agreements point by point. In the discussions on the piece wage paragraph both sides motivated certain of their proposals. Discussions on the piece wage excess were held back to be dealt with at the same time as wage questions.

Paues pointed out that the lack of flexibility in the fixed piece wage schedules had created difficulties during the period of the agreement being terminated. This had caused the employers to suggest that piece wage schedules of a shorter validity could also be arranged. As justification for the Textile Employers Association's proposal that in piece wage disputes a transfer should be made to hourly pay at once, Paues stated that the system could speed up an agreement.

The spokesman for the workers justified the union's proposal that the agreement on piece wages should be made between the employer and the local union with the argument that the piece wage should be decided between equal parties. It was necessary that on this point the unions should protect its members' interests more actively.

There was no more detailed discussion on the Textile Workers Union's proposal on paragraph 15. Finally the two sides agreed to continue negotiations on Friday, 19 November.[21]

The day before negotiations were to be resumed the board of the Textile Employers Association held a meeting. The first round of negotiations had called for another look at the draft agreement and Erik Paues now put forward a list of changes in the employers' draft. This was approved by the board and so it was possible to present it to the other side at the negotiation contacts on the following day. The alterations did not affect the piece wage paragraph where the employers' position remained unchanged.[22]

During the three days from 19–21 November the two negotiating teams continued to discuss the proposed agreements point by point in those parts which concerned the general provisions and the joint rules of procedure. The employers now took up the rationalization provision. It had shown itself to be very difficult, stated Paues, to carry out rationalization measures in the factories because of the workers' compact opposition to scrapping the fixed piece wage agreements and making new ones. The cause of this was that the old agreements gave an excess above that laid down and the workers wanted to protect this building up. The employers, however, could not agree to being deprived of the possibility of introducing new piece wages if changes occurred in the conditions of these piece wages. If the employer could show that a new piece wage offered provided a piece wage excess in accordance with the agreement it was not possible for the workers to refuse to accept the offer.

For the Textile Workers Union it was emphasized that the current provision could not be changed. A change might have the effect that current piece wage earnings could be reduced at any time. Nor was it possible to accept a demand for a transfer to hourly pay in piece wage disputes. It was thought that it was necessary for negotiations to be carried on by two equal parties, that is to say between the main organizations, before any change in the earnings position could be tolerated.

With regard to the negotiation procedure the workers had made proposals, partly that piece wage agreements should be made between management and the union, and partly that all suggestions about piece wage agreements should be made in writing in order to avoid disputes, and further that if either party so requested the negotiations about piece wage agreements should be initiated within 14 days.

Erik Paues confirmed that a ban on companies concluding piece wage agreements directly with the workers was unacceptable to the Textile Employers Association. He considered that the other proposals would in practice prove themselves to be impracticable. No discussion on the proposed provision from the workers' side about production investigation is shown in the available minutes for that round of negotiations.

The last day of this round of negotiations, 22 November, was devoted to the wage question and it was soon discovered that the two proposals were unreconcilable. Nor in the negotiations that followed immediately afterwards between representatives of the Textile Employers Association and the Textile Workers Union would either of the parties make the concessions demanded. The negotiating teams decided to terminate the negotiations and to request the Social Welfare Board to appoint a conciliator before whom both parties would be prepared to resume discussions at the beginning of December.[23]

The first two rounds of negotiations during November 1930 were taken up by discussions about the proposals for alterations in the general provisions of the agreement, which both parties had presented. Both of the organisations' main lines in the piece wage question seemed to be quite clear. The employers were trying to get a local flexibility in the piece wage system that would permit a smooth adaptation of wage costs to those considered necessary by management. At the same time a provision should be introduced into the national agreement that in reality was to function as a ceiling for piece wage earnings and thus as a barrier to wage drift beyond control of the association.

The workers aimed at such changes in the piece wage paragraph that would even out the local balance of power between the parties who made the piece wage agreements. This could only occur by giving the organization a say in piece wage agreements. In addition the workers' representatives kept an eye on the provisions of the agreement that seemed suitable to ensure piece wage volume and the level of earnings.

Apparently the views held by both sides about the general provisions were widely separated. When in addition it appeared clear that on the pay question they were also obviously on a collision course both parties preferred to stop the negotiations and bring in a neutral chairman. With this the negotiations entered a new phase.

4. Before the conciliator

At the end of November the government appointed Allan Cederborg as conciliator. He had had a long experience as a leader of negotiations. Cederborg called for a renewal of negotiations on 2 December. This round too started off with a discussion on the general provisions of the agreement and the parties presented their proposals and views. In connection with this the employers also took up the question of piece wage earnings. In the document presenting notice of termination by the Textile Employers Association the demand had already been made for a general revision of piece wages so that they became in line with the agreement.[24] Erik Paues now pointed out that the differences in the piece wage excess were considerably even within one and the same plant. They varied between a minimum of under 15% and a maximum of 80% of the hourly wage. The Textile Employers Association retained the right, Paues confirmed, in connection with a new agreement to so regulate piece wages that the excess came into line with the agreement. In reply to this K. J. Olsson stated that the workers could not agree to any revision of piece wages that resulted in a piece wage excess below that of the 30% average requested by the workers.

So after again establishing that the view on the general provisions were incompatible the wage question was taken up. Not here either was it possible to discern any unanimity so it was decided to adjourn the negotiations until 10 December.[25]

Not even in the round of the negotiations started on 2 December had the

Table 15. *Textile industry. Statistics reported to Swedish Textile Employers Association on piece wage conditions: number of hours worked and average piece wage excess as a percentage of hourly wages*

Source: Statistical information in Textile Employers Association's negotiation material 98/1930

	Male workers			Female workers		
Branch	No. of plants	No. of working hours (1 000s)	Average piece wage excess	No. of plants	No. of working hours (1 000s)	Average piece wage excess
Whole textile industry	104	6 439	34.3	126	25 388	37.9
Wool	30	2 215	30.7	42	6.316	34.9
Cotton	34	2 714	28.0	38	10 843	33.3
Knitwear	21	604	44.3	25	4 314	45.9
Jute	2	45	15.8	2	1 305	28.5
Hemp	4	145	36.0	4	397	30.3
Linen	5	337	40.1	5	1 373	45.2
Ribbons	3	85	41.7	3	300	55.7
Dye works	2	194	29.0	3	91	40.8
Various	3	100	52.1	4	449	49.6

parties approached each other to any noticeable degree. Instead a further issue was introduced; the principles for a general revision of piece wages.

Although this revision of piece wages was brought into the negotiations at such a late stage it was in fact here that the most important point lay for the employers, the inflexible demand. Already at the board's meeting prior to the second round of negotiations in November, Axel Bergengren had declared that he was prepared to relinquish all changes in the general provisions and reductions in the hourly wages if only such a general revision of piece wages could be brought about that reduced the excess to that average of 15% guaranteed in the agreement.[26]

Obviously there were great gains to be made here. A compilation of the piece wage excess at all plants in 1929 shows that the spread was considerable and that the average for the whole industry, as shown in Table 16, lay clearly above the agreement's target value of 15%. A reduction to an average of 15% above the hourly pay would decrease the costs of piece wages to the employers by roughly 3.5 million Skr per year.[27]

Following the negotiations at the beginning of December the general council of the Textile Employers Association adopted Bergengren's assessment as its own. It then put forward as its minimum demands an agreement based on unchanged hourly wages and a general revision of piece wages. If this was impossible to achieve at the negotiations a general lockout should be used. The council decided to request permission for such offensive action from the Swedish Employers Confederation.[28]

Negotiation was resumed on 10 December when Cederborg met each party separately. The next day there were private discussions and on the 12th the parties were able to define their points of view and report changes in their original proposals. On the basis of these Cederborg prepared a memorandum, a feeler. An answer to this was requested by the following day. From this it was clear that it was still impossible to reach any agreement. Negotiations were adjourned again until 29 December.[29]

The written offers exchanged and the answers to them provide a possibility of showing the role played by the peice wage question at that stage of the negotiations with a good deal of accuracy.[30]

The stand taken by the Textile Employers Association on the piece wage question following the private discussions of 11 December had not softened at all, rather the reverse. The employers still demanded the right to a general revision of piece wages. In addition they wanted, as was put forward in the discussions at the middle of November, to add a minute to the agreement signifying that a worker had no right to refuse a piece wage that gave the excess laid down.[31] In their reaction to this offer the workers stated that their own demands remained unchanged. A provision that would shackle the workers' freedom to refuse a piece wage offer would not be accepted in any circumstances.[32]

The result of the union's private discussions on 11 December did not bring the parties closer together either. Demands were still made for increases in hourly pay of 10% (15% for women's wages), and piece wage compensation of 10 öre/hour (7 öre on women's wages). In addition it was demanded that the local unions should get joint determination in a number of questions including agreements on all piece wages.[33]

Guided by the offers he had received the conciliator, as already noted, put out a feeler to get both parties to state their positions more exactly. Cederborg suggested that they should conclude a one-year agreement, which was to include the changes already agreed and which were in the other points identically worded with the old agreement. The hourly pay should be left unchanged and the revision of piece wages done in such a manner that those lying under the 15% excess should be increased to that level whilst higher piece wages should be cut down to the limit of a 35% piece wage excess, "but not further than that the overall total of the new piece wages for every workplace could be calculated to provide the same amount of excess as the old overall total of piece wages had provided." Finally the conciliator suggested that consultation between management and unions should occur on a number of questions, working time and the dates of holidays, "or other questions of importance for the flow of work". Nor should discharge occur without consultation. At this consultation the management's reasons for the measure should be stated and the person to be discharged heard.[34]

Cederborg thus fastened on the employers' strong demand for a revision of piece wages and the workers' demand for increased joint determination even though it was mentioned that this joint determination should also apply to piece wage agreements.

In its reply to the feeler the Textile Employers' Association rejected the idea that the workers should get increased influence. It maintained the demand for a thoroughgoing revision of the general provisions and for a general revision of piece wages. But the employers were more amenable on the demand for cutting hourly wages. The wages must, it was said, lie between the status quo and the claim of the employers.[35]

Nor were the workers prepared to retain the provisions of the old agreement on points where there was disagreement. On hourly wages the Textile Workers Union kept firmly in principle to the claim for increases but smaller adjustments in the stated figures could be considered. Finally, piece wages could be revised so that the upper limit became 35 % if the minimum limit was given as 25 % above the hourly wage.[36]

After these statements it was clear that continued negotiations were hopeless unless something occurred that caused either party to reconsider its position. As two weeks remained before the agreement ran out it was decided to wait and see and to meet again for discussions on 29 December.[37]

Both parties were fully aware that the chances of achieving any result were small. The Textile Employers Association's general council laid down its line of negotiation as early as 14 December. If negotiations broke down at the end of the year all members of the association should lower all hourly wages from 7 January by 10 %. Also all piece wage work should be carried out at this reduced hourly wage until more accurate production investigations had been carried out and new piece wage rates confirmed. Following this the piece wage earnings should be paid retroactively. Otherwise the provisions in the old agreement should be used.[38] The decision concerning piece wage earnings was completed later so that the piece wage excess was generally decided at 15 % above the hourly wages reduced by 10 %.[39]

The available material does not show that any similar activity was started by the other side. At the meeting of the executive board of the Textile Workers' Union on 18 December K. J. Olsson reported on the negotiations. The report was approved after a short discussion but no course of action for what was to occur is shown.[40]

On Monday 29 December the parties met once again under the chairmanship of Cederborg.[41] As expected both Paues and K. J. Olsson confirmed that nothing had occurred that gave them any reason to change their attitudes. Cederborg reported that he had informed the government about what had taken place and that the Minister of Social Affairs wished to meet both parties. Negotiations were therefore broken off for a couple of hours whilst Paues and Olsson individually met the Minister, Sam Larsson, a liberal. The visits were perhaps a formality but at least a sign that the position was regarded with increasingly greater seriousness by government; the threat of conflict was increasing.

However, private discussions were resumed and Cederborg made a new attempt to reconcile both parties. On the very same day he presented a new memorandum that largely followed the feeler of 12 December. Hourly wages should remain unchanged and piece wages so revised that they came within the limits of 20–35 %. The point about joint determination from the early memorandum also remained. A small joint committee should go through all the general provisions of the agreement and within two weeks present a proposal for the decisions of the negotiating teams. Where opinions differed the agreement should follow the old one. During this period of two weeks the non-strike rule would apply.

A new clause was introduced in this connection with provisions about how the basis of setting piece wages should be constructed. Cederborg suggested that the following minute should be approved:

> The parties are further agreed on the minute that where questions arise in the future about the setting of new piece wages or the revision of all such wages, the preparatory testing should occur in such a manner that for this purpose specially fast or skilled workers are not selected nor an unexperienced or slow worker, but that the testing during the time required takes place with workers of average capability.

If testing had taken place in accordance with this provision and if the piece wage thus calculated lay within the stated limits for the piece wage excess, the workers would be unable to refuse the piece wage offered.[42]

Cederborg thus tried the idea of separating the wage questions from the general provisions and initially to obtain direct agreement on the former, linked only to certain points in the latter. The grounds for calculation of piece wages, how the production investigations should be carried out, were taken up for the first time in the active work of negotiation as a counterweight to the employers' demand that the workers should accept the piece wage price offered if it fell within the limit-values of the piece wage excess.

The workers did not accept this separation but demanded immediate agreement at the same time on wages and the general provisions. They maintained their demand for a 10 % increase of hourly wages but were prepared to accept a revision of piece wages to the limits of 20 and 35 % if the provision was included in the agreement that piece wages should give on average 20–35 %. Regulation of how the production investigation was to be carried out was approved but the obligation to accept a piece wage was definitely declined.[43]

The Textile Employers Association accepted the status quo for hourly wages on condition that a revision of piece wages was made to limits of 15 and 25 % respectively. Regulation of the production investigation was rejected by the association, which recalled the provision in the current agreement laying down that piece wage agreements should be made between the employer and the workers concerned. The association thus defended the *freedom* of piece wage agreements that they considered excluded every provision as to how piece wages should be offered. However, a provision could be accepted that as a basis for the production investigation should lie the *average* output of a plant's workers of normal skill if the number of workers was sufficiently large.

The responsibility for accepting the piece wage offered was greeted by the employers with enthusiasm and they wanted to have a completely compulsory wording. The sweeping phrasing about joint determination was rejected with reference to the employers' right to lead and allocate the work, and to engage and discharge workers.[44]

Both parties' reactions to Cederborg's memorandum showed that they were still far apart. Nevertheless he made a final attempt in a suggestion to both parties, a final proposal that they had to accept or reject.

Cederborg's suggestion presumed unchanged hourly wages in accordance with the employers' offer. The workers were recompensed instead with a proposal for piece wage compensation with 8% of the hourly wage, which meant such increase in hourly wages as did not affect the piece wages calculated as a percentage of the hourly wage. Piece wage revision was suggested restricted to the limits of 15 and 35% respectively, a compromise where the employers' lower, and the workers' higher, limits were used. With regard to the production investigation the formulation was proposed that the piece wage "so far as possible is calculated with regard to the performance of a worker of normal skill and under normal working conditions". To have this provision was thus to meet the workers halfway and the employers would certainly have opposed it. As a quid pro quo Cederborg maintained what was laid down about the obligation to accept such a piece wage as had been calculated by means of a production investigation correctly undertaken and that showed it provided an excess within the limits established. The point about joint determination from the previous memorandum was deleted under the pressure of the firm opposition of the employers. The workers' demand for connected agreement, on the other hand, received a direct breakthrough in Cederborg's suggestion.[45] The employers accepted on the whole the proposed hourly wages but definitely rejected the piece wage compensation. With regard to the revision of piece wages they stayed firm on the limits of 15–25%. The proposal concerning investigation of production and the obligation to accept piece wages was accepted by them.[46]

Again the Textile Workers Union accepted the status quo in respect of men's hourly wages but persevered with a demand for a 5% increase in women's wages. Furthermore the union requested a 10% piece wage compensation instead of the proposed 8%. The demand for a piece wage revision in the earnings area between 20–25% to be written into the agreement was maintained as was the demand that the obligation to accept a piece wage should disappear.[47]

On 30 December 1930 the parties were still widely separated in important respects. The employers were prepared to accept unchanged hourly wages whilst their opponents, in the shape of compensation for piece wages, still demanded that they should satisfy their low wage group, the workers on hourly wages, and in addition through an increase in wages for the women to decrease the inequalities between the sexes. With regard to the piece wage revision question the principle manner of procedure was agreed, but there was a division of opinion about the limit values and their meaning. The employers thought that low or newly set piece wages should be increased up to 15 % and high piece wages should be written down to the 25 % level while the remainder should be left untouched. The workers wanted to see the values of 20–35 % above the hourly wage as sector within which earnings should lie. This meant that the excess of newly set piece wages according to the agreement should take a value within this area, not necessarily just at the lower level.

Faced by these differences in ideas the conciliator had no other choice than to declare the negotiations broken off and the two parties free to take those steps they considered suitable.[48] This also automatically erased all those consents and undertakings given by the parties during the negotiations. The conciliator was also to continue to follow the development of the parties' relations with each other and requested that they should let him know if anything occurred to change the situation. They then went their different ways.[49]

The phase now completed in the negotiations for a new textile agreement had been led by a neutral chairman. It seems to be quite clear that Cederborg's contribution, although it did not lead to the signing of an agreement, provided a contour not only to the agreement that was to come but also had consequences for the construction of the textile agreement for many years.

The employers stand out as the stronger party. Their main aims, status quo for hourly wages and the reduction of piece wages, were in principle accepted by the workers. With the assistance of the conciliator, however, the construction with a sector laid down in the agreement for the piece wage excess had been introduced into the discussions.

As the reduction of piece wages had been accepted in principle even though the limits were still disputed, interest was directed to the kind of revision and the new setting of piece wages. There the workers earlier demands about the investigation of production provided a suitable token to link up with the wish of the employers to introduce an obligation for the workers to accept piece wages within the area of earnings laid down in the agreement. It was thus in a negotiation situation subjected to pressures characterized by the efforts of the conciliator to discover firm ground on which to build agreement between the parties, that the provisions about the bases for the calculation of piece wages were introduced into the negotiations on the textile agreement.

The new line of thought manifested in this way did not suffice however to create conditions for a new agreement. Collision was inevitable.

5. The conflict

At the end of the year when the current agreement between the Textile Employers Association and the Textile Workers Union ran out, the position was thus as follows. Negotiations between the parties had broken down. The employers had their plan ready, by a unilateral decision to cut wages by 10% and to stop piece wages until more accurate production investigations had been carried out on which new piece wages could be fixed in order to provide a 15% excess.

No decision had been reached yet by the workers. On the first day of the new year the union's representative council met once again for discussions with its own negotiating team. K. J. Olsson reported on the negotiations up to the time they were broken off and carried his reasoning up to a presentation of three alternative lines of action: (1) An immediate strike over the whole line, (2) a partial strike or (3) no strike until the employers had started their offensive measures. He thought the first alternative was best if the financial resources were good. If this was not the case the second was preferable. The third alternative was good if the employers were prepared to let work proceed for a longer time as this would enable the union's coffers to be filled, but if it was only a case of gaining a few weeks there was not much point in it.

K. J. Olsson had been in favour of an immediate general strike but a conversation with the chairman of the Trade Union Confederation, Edward Johansson, who recommended the second alternative, had caused him to understand that it would be a good idea to try this more careful path, say the minutes. It should be remembered here that the Textile Workers Union did not belong to the Trade Union Confederation. The information that Olsson received from the confederation's chairman was without doubt that the textile workers could not count on any great financial help from the Trade Union Confederation, especially in the case of union offensive. The union would have to rely on their own resources. By selecting alternative (2) the eventual responsibility for escalation of the conflict could be placed on the employers.

But what principles should be used when the partial strike was carried out? Either one could take certain categories of workers out on strike at once or certain companies could be selected. The latter was preferable and in this case the large companies should be considered first.[50] After a certain amount of discussion the representative council agreed to recommend the careful path of a partial strike.

At a meeting on the following day the executive board followed the recommendation of the representative council. Seven large companies were selected and at these, the executive board informed the other side by telegram, work was to be stopped on 7 January if no agreement had been reached before then.[51] The list presented consisted of large and leading companies in the Textile Employers Association, which in itself meant financial pressure on the employers' financial resources. Altogether the seven companies employed approximately 8 000 workers.[52] It should also be observed that all the large branches were represented: cotton, wool, linen and knitted goods, and that the leaders of all of the companies selected (with the exception of Nordiska Maskinfilt) were members of the Textile Employers Association's general council. Thus things were arranged in such a manner that the pressures of the strike became directly felt for those in the employers' camp who had to decide on further measures.

Officially the workers had taken the first step in escalation of the conflict and this put the employers in a favourable position. The general council of the Textile Employers Association had already announced internally its intention to cut wages and had had posters printed to be put up in the plants on 3 January. At a meeting immediately following the breakdown in negotiations on 30 December the council had confirmed this decision.[53] Information was sent out to the members on the last day of the year.[54] But as the workers were not to be informed of the measures until 3 January the Textile Employers Association received the telegram from the union before this. This made it possible for the employers to resent their long-decided measure to reduce wages as a countermove against the decision of the Textile Workers Union for a partial strike.[55] Thus the employers made a gain in public opinion. The Textile Employers Association informed their opponents that because of the workers' strike action it had instructed its members that from and including 12 January they should apply a 10% cut in hourly wages and piece wages that were regulated at 15% above the decreased hourly wages.[56]

This communication from the Textile Employers Association immediately caused a new meeting of the executive board of the textile workers. Again the discussion was about three alternative modes of action: (1) a general strike at the factories belonging to members of the Textile Employers Association, (2) a strike at the factories where reductions were actually made and (3) a continuation of work but a recommendation to the members to go slow and refuse overtime and shift work. No decision was arrived at immediately but the question was referred to a new meeting two days later.[57] It can be presumed that these days were also utilized in investigating what support the textile workers could get. When the discussion was resumed on 5 January it was clear that both the Loom Tacklers Union and the Machinists, Stokers and Repairers Union were at one with the textile workers.[58] After some consideration the board decided for alternative (2), to strike at the factories that followed the directive on wage cuts and to let the work continue at the employers who did not obey the message from the Textile Employers Association.[59]

A number of reasons can be presumed to lie behind the workers' choice of line. To a certain extent it was a speculation in the lack of solidarity among their opponents; a test of how strong their unity was. Also it was a way of repainting the picture of the employers so that public opinion regarded the employers as the aggressors. The workers did not extend the conflict but left it to the individual employer to assume responsibility for a reduction in pay and the consequences this entailed. The executive board thus instructed its members that from and including 12 January they should cease work at the factories that from that day applied the cuts in pay.[60]

The Employers Association's only reply to this new move was through a circular to its member to emphasize the importance of internal unity and the impossiblity of making any general exceptions from the decree issued.[61]

In view of the overhanging threat of conflict Olsson and Paues were again summoned to the Minister of Social Affairs, when it could only be noted that the chances of rapid solution of the differences were nonexistent. Therefore all parties were agreed that at that time no conciliation commission should be set up.[62]

The great textile industry conflict thus became a fact. Speculation about disagreement among the members of the Textile Workers Union, if such speculation was really seriously meant, was completely wrong. On 12 January there was a complete stoppage of activities in all plants belonging to members of the Employers Association with the exception of seven. Two of these had agreements with other unions that applied up to the end of 1931. These factories had therefore been excepted from the pay cuts. Of the remaining five the workers, who were unorganized or organized in other unions, had accepted the lower wages and continued to work. At one of them they had later, on 21 January, gone on strike. So even from the Textile Workers Union's point of view the internal solidarity was strong: the workers had obeyed the order of their leaders.[63]

After a week the conflict had reached the extent that persisted until the middle of March, when 116 employers and 143 workplaces were affected. Approximately 32 000 workers struck, some 9 000 of these being unorganized.[64]

Quite obviously the conflict became expensive for both parties but perhaps most for the workers. The deteriorating trade situation during the winter of 1930/1931 worked in favour of the employers. The Textile Workers Union's own financial resources were rapidly consumed. By the middle of March the amount of strike pay was reduced.[65] Altogether the conflict cost the union about 2.6 million Skr. As the union was outside the Trade Union Confederation the possibilities of regular support were limited.[66] In spite of this contributions of about 200 000 Skr flowed into the strike fund during the course of the conflict.[67]

The strike was called off after a settlement on a new agreement was reached on 20 March but not until a week into April was work resumed completely.[68]

The extensive textile conflict was a natural result of a very locked position in the negotiations about the textile agreement. Neither party wanted a conflict but both were prepared to accept it. As early as the policy conference on the agreement in September K. J. Olsson had foreseen that notice of termination with such extensive demands as were suggested could result in a conflict (see p. 164). When the course was set for collision it was obviously up to the parties to react so that the opponent seemed to be the agressor. This was important not least because of the rules for assistance of the Trade Union Confederation. Here the employers were more successful. Without any actual conflict measures, by cutting wages in the new year when there was no agreement in existence the Textile Workers Union was forced off the careful line it had selected and on to the considerably more risky alternative, that of a total union strike. The resultant development of the conflict and the negotiations were to make clear that the workers had aimed too high.

6. Solution by negotiation

When the textile people had been supping the bitterness of the conflict for a few weeks the conciliator, still Allan Cederborg, renewed contacts with both parties, contacts which gave him reason to believe that the parties could be brought to negotiation again at the beginning of February.[69] On 6 February the government appointed a conciliation commission for the textile conflict. Allan Cederborg became chairman and so the continuity between the earlier and the later rounds of negotiation was good. Other members of the commission were the conciliators Hilding Kjellman and Olle Ekblom.[70]

The conciliation commission summoned the parties to negotiations on 12 February with a continuation on the following day.[71] The position still seemed to be locked.[72] Cederborg's proposal from December about smaller negotiating teams was then taken up and a subcommittee from the parties appointed to go through the provisions of the agreement once again, point by point.[73] There were two more discussions, one at the middle and one at the end of February. On the basis of the subcommittee's work on 2 March the commission presented a draft at the request of the parties. A reply was requested by 5 March.

In the subcommittee too the work seems to have gone rather slowly.[74] The negotiating team of the workers reported to the union's executive board and put the blame on their opponents' obstinate and continued demands that benefits already approved should be deleted from the agreement.[75] It seems obvious that this applied not least to the rationalization clause where of course the employers wanted to have a considerable modification in the text of the agreement. The attack also seems to have been directed against the higher remuneration that was paid for especially expert or heavy work.

The commission's draft agreement of 2 March contained the same limit-values for the original piece wages earlier proposed by Cederborg, 15 and 35 %. The piece wages target was stated in line with the workers' demand for *at least* (not 'on average') 15 % above the hourly wages. The provisions about production investigation had been included exactly as the workers' claim. To balance this the earlier proposal about piece wage compensation had been deleted.[76] In its reaction to the draft the workers noted the positive characteristics of the proposal but maintained their demands for piece wage revision to the limit values of 20 and 35% and a contractual piece wage excess of 20%. In addition the demand was still made that piece wage agreements should be concluded between management and the trade-union.[77] Following further contacts with the commission, two days later the workers clarified their most important views in a memorandum. But this contained no changes with regard to the matters dealt with here.[78]

The employers were also critical to the proposal and on the whole maintained their demands; a reduction in hourly pay of up to 10%, revision of piece wages to the area of 15–25%, the target of piece wages to be on average 15%, a minute to the effect that piece wage schedules contrary to the agreement should not be approved and a more flexible wording of the rationalization provision. They were also completely negative to the proposal to introduce provisions about production investigation, not least with reference to the possibility of carrying out time and motion studies. Finally the employers still demanded that the obligation to accept piece wages that gave a contractual excess should be included in the agreement.[79]

The negotiations did not seem to be getting anywhere. The conciliation commission therefore decided to bring the round of negotiations to a close by making a formal request to the parties containing a proposal for agreement, which only differed on one point from the draft on the matters dealt with here. A provision had been added that provided opportunity for the adjustment of the relations between piece wages in individual companies. To this request, which had not been established with the parties, the commission requested an answer by 16 March.[80]

The general council of the Textile Employers Association considered this final proposal from the conciliation commission at a meeting on 9 March. With regard to the revision of piece wages and the piece wage paragraph the council again confirmed the demands that had already been but forward because of the draft.[81] In addition the council wanted to limit the higher, special piece wage excess to 30%. Guidelines were also drawn up for the revision of piece wages as follows.

The implementation of the piece wage revision out in the plants would be carefully followed up by the association. Before new price schedules were confirmed the workers would be on hourly pay according to the schedule and production investigations would be carried out. Following this the piece wage schedules would be confirmed and the workers receive piece wage earnings retroactively. With regard to the possibly higher earnings (from 25 to 30% according to the employers' proposal), a questionnaire investigation would be carried out among the association's members about the continued existence of such benefits. A final answer could not be delivered to the commission until this had been carried out.[82]

How the Textile Employers Association obtained the views of the members is not known. When on 16 March the association presented its reply to the conciliation commission the question of the extra-high earnings was taken up in some introductory consideration. The association stated that the commission's proposal was mainly based on the status quo, which the association had stated it was prepared to accept in the main although the industry during the past decade had been faced by greatly deteriorated conditions on the market. It was important also that possibilities of changes in production were ensured. At this point the association's communication contained a formulation that clearly showed how the revision of piece wages was regarded, not only as a practical but also as a principle question:

With regard to the establishment of the new piece wage schedules the commission's proposal provides certain possibilities to undertake the revision of older piece wages and in connection therewith to arrange working conditions on a rational basis. What on the other hand refers to working conditions otherwise in the plants, no such possibility is provided. On the contrary it appears in accordance with point 5 in the conciliation commission's proposal that the constraint of the agreement henceforth shall be placed on such wages and limits, which the employer has freely given over and above what is laid down in the agreement or that have crept in through an erroneous application of the agreement. It must be considered as highly inconsistent to except these wages and benefits from revision when, with regard to the other provisions, changes are granted . . .

Here the employers were taking care of their *freedom of action*, freedom to alter the conditions of production and thus make operations more economical without the interference of other parties or being tied down by older piece wages. In this freedom of action lay the principal idea, with a strong emphasis on the revision of piece wages. The other parts of the employers' reply may be interpreted as a logic extension of this. They were thus not wholly prepared to base the revision of piece wages on earnings during the period proposed (1.10 1929–1.10 1930) but wanted to retain a certain freedom of action. The demand to remove the proposed provisions on production investigation was also justified with similar arguments;

We request that this point be deleted from the agreement as it only deals with one of the many ways by which the amount of production per unit of time can be calculated. Further, the retention of the provision can easily lead to a misunderstanding that, for example, time and motion studies and other evaluations of piece wages may not be carried out.[83]

For the workers the results of the negotiations during the first week of March were reported at a meeting on 9 March of the executive board of the Textile Workers Union, which decided to call in the representative council on the following day.[84] At its meeting in the afternoon of 9 March the negotiating team was fully aware that time was working against its interests. The finances of the union were taking a thrashing. The assessment of the conciliation commission that the proposal submitted represented the last way of clearing up the conflict rapidly was taken ad notam. The negotiating team proposed that the offer should be accepted. The following day at its meeting a majority of the representative council shared the assessment of the negotiating team. It was decided, however, that the proposal of the conciliation commission should be submitted to the members' scrutiny in a ballot.[85]

The circular about this ballot, which was sent to the branches of the union, left no doubt in its first sentences about the position taken up by the representative council and the negotiating team. It was made clear in the text that the wage cuts and the other changes for the worse that the employers had demanded had not penetrated into the commission's draft. Only with regard to the revision of piece wages did the proposal mean in certain cases a reduction to the 35% excess. But this disadvantage was balanced however, it was said, by "certain important improvements". Among these was noted that all piece wages should give at least a 15% excess. Certain of the workers' demands had not been met it was admitted, including the demand for piece wage compensation for those workers paid by the hour. Decisive for the recommendation to accept the proposal now before them was, however, the strained state of the union's finances, which did not permit any continuation of the struggle.[86]

The ballot lists were returnable at latest by 15 March. The result was a marked defeat for the leadership. The proposal of the conciliation commission was rejected by a large majority.[87]

The executive board had appointed a new negotiating team, which guided by the result of the ballot was to deliver an answer to the commission. The great defeat for the line taken by the representative council obviously made the position more difficult.[88] The representative council was therefore summoned again for 16 March, the same day the reply was to be handed over. Finally after a long day's meeting it was possible to deliver the reply to the commission, seven hours late.

In the reply the Textile Workers Union stated that the conciliation commission's proposal had been rejected by the members with a large majority. The main complaints were that the piece wage compensation of 10% to the hourly paid workers had been deleted and that four districts had been moved to a lower schedule. In addition there were a number of complaints about the draft of a more specific character.[89]

To summarize the situation on Monday evening 16 March 1931, the following can be said. Both parties had, although with somewhat differently formulated questions, approached their members about the proposal, which had in this way been discussed most widely among concerned. Neither had wholly accepted the proposal although both had accepted its main lines. But each party stood fast on certain demands, essentially one offensive and one defensive demand from each. If the wishes of the workers about piece wage compensation (offensive demand) and their defence against schedule reductions (defensive demand) were satisfied, the union would certainly allow its other eight criticisms to fall. It was also clear that the employers were prepared to accept the draft if only their principles about the grounds for the revision of piece wages and the carrying out of this were accepted, and if the provision on production investigation was left out of the agreement. Apparently the conciliation commission considered that the proposal had been received so positively that this provided reason for continuing the negotiations in spite of the depressing result of the ballot among the workers. So during the days that followed the commission entered into private negotiations with both parties on the basis of the rejected proposal.

The commission decided first of all to go to work on the demands of the Textile Workers Union. At its meeting on the 16th, the representative council had again emphasized that it was the hourly pay increase in the shape of piece wage compensation that was the most important demand.[90] The commission was able to find an opening for continued conciliation in a personal letter K. J. Olsson had sent to Cederborg as early as 8 March. There the union representative stated certain demands from which the workers could not depart, "in the case they can be persuaded at all to give up the demand for special bonuses for hourly paid workers".[91] Among these demands there were some considered impossible by Cederborg, others that were already satisfied in the draft agreement, and still others that were apparently negotiable.[92] The most important of these concerned the moving of Borås to a higher schedule; this should mean an increase in hourly wages for a good deal of the union members and retention in the original schedule of the districts proposed for moving downwards.

During the afternoon of 17 March the conciliation commission met the workers' negotiators and proposed that no districts should be moved down (Cederborg did not yet mention the case of Borås) and that the holiday entitlement for the current year should be left untouched by the conflict if the workers accepted an immediate agreement. An answer was to be given on the following day.[93]

On the evening of the 17 March the commission met the employers and presented the same proposal to them. What was interesting for the Textile Employers Association was the possibility of getting an agreement across the table. The workers' negotiating team had of course shown themselves to be more accommodating than the majority of their members. But in the proposal to the employers Cederborg added to the mixture the promotion of Borås, which he had not yet mentioned to the other party. If the employers accepted this construction Cederborg had thus created quite a large capital for negotiation that he could place directly on the table at continued negotiations with the workers.[94]

Paues replied on 18 March: if the workers gave up the demand for piece wage compensation and agreed to a settlement across the table the employers were prepared to accept the construction presented by the commission. By using the promotion of Borås the conciliators were able during the 18 March to persuade the workers also to accept the proposed settlement. To Hilding Kjellman, who was at home in Gothenburg, the following triumphant telegram was sent:

A child is born and known in spite of mutter
machines again go putter putter—
commission cheers huzzah huzzah.
Allan Olle[95]

On 19 March the new agreement was signed by the Textile Employers Association, its opposite numbers the Textile Workers Union, the Loom Fixers Union and the Amalgamated Unions. The validity of the agreement however was made dependent on signing by the Stokers, Machinists and Repairmen's Union. The necessary signature was appended on the following day when the Employers Confederation also endorsed the agreement.[96]

The settlement meant that all hostile measures should cease and that work should be resumed as soon as possible. Revision of piece wages to the limits 15–35%, with continued regard for special qualification and degrees of difficulty, was to be started immediately and result in new piece wage schedules that were to be ready within a month. The revision was to take a long time and covered 124 plants.[97]

7. Role of the trade union opposition

In summing up the collective negotiations 1930–31, the most dramatic in the history of the textile industry, the employers could claim they had won the day. The workers' offensive had been parried and the main demand had been satisfied, that for a radical revision of current piece wages, even though the limits had finally been set more generously than had originally been thought, which reduced the immediate saving. More important was the possibility, with the application of new methods of calculation, immediately to examine the basis of piece wages. In these production investigations the employers had certainly been compelled to allow the workers some degree of joint determination. It was, however, one of the concessions required by the situation, and as soon as the chance arose the employers were again prepared to tear up the whole regulation of the basis of piece wages. The fact that they had succeeded in forcing the workers to an agreement across the board could also be regarded as a practical gain. This was a blow to the ballotting procedure, which the employers regarded as a great hindrance to rational wage negotiations. The fact that this concession was forced quite contrary to the ballotting of the members already carried out certainly underlined the strength of the employers' position but at the same time created difficulties for the reformist leadership of the Textile Workers Union, which on its merits was presumably not an effect desired by the Textile Employers Association. Also as a success in the long-term from the employers' point of view can be regarded the weakened finances of the workers, which guaranteed industrial peace for at least the next few years.

The guiding principle of the employers' actions during the collective negotiations 1930/31 was, however, undoubtedly the intention of safeguarding or increasing the freedom of action for management in relation to the workers. The revision of piece wages in conjunction with the new agreement was the instantaneous expression of the need felt by the employers to adapt the whole of the piece wage volume to such changed conditions as were not covered by the rationalization clause. But with even such periodic adaptations the piece wages tended to freeze and therefore the employers tried to increase the flexibility in the system—freedom of action of the stronger—by means of modification of the rationalization operations and by preventing the regulation of production investigations. The same purpose was also served by defending the freedom of piece wages in the sense that even in the future piece wage agreements should be concluded directly between management and the workers affected without the interference of the trade union organizations.

Seen from the other side of the hill the result of the conflict naturally seemed more gloomy to the Textile Workers Union. The demands with which they had entered the collective negotiations were far from being fulfilled. An increase in hourly wages or compensation for piece wages, like the six day holiday, had been conspicuous by their absence. The demand for a contractual piece wage excess of on average 30% had resulted in a provision giving at least 15%, a change for the worse but not completely without advantages. With regard to the piece wage paragraph and equality between management and workers the union had succeeded in pushing forward its positions through changes in the wording of paragraph 7 clause 1b and through the provision on production investigations. The last was also considered to prevent the employers from using other calculations for the basis of piece wages than the comparatively simple statistical investigations of actual production. On the other hand the textile workers did not obtain a provision that free piece wage agreements should be concluded between the employer and the trade union. As well as the original offensive demands thus being frustrated the workers were compelled to accept the original piece wages that meant a direct cut in the piece wages. However, the idea that greater skill and experience should constitute a basis for higher pay had been successfully defended by the workers. In spite of these lesser successes the Textile Workers Union's leadership could not hide the fact that in spite of staking the whole of the union's resources it had not succeeded in improving the wage level that had remained practically unchanged since the beginning of the 1920s.

It seems almost obvious that the collective negotiations from their conception at the policy conference and up to the settlement, increased the differences between the more offensive and more restraining opinions in the Textile Workers Union. Already when the decision to give notice of termination was made the leadership met defeat against a more radical opposition. In this context it is impossible to decide how this opposition was made up, what proportion consisted of generally radical union representatives and how many conducted their union actions more directly from political ideologies and political directives.

During the policy conference to discuss the agreement the communist John Berglund of Karlstad and the spokesman for the Stockholm branch, Sven Dahlström, were responsible for the most violent attacks on the leadership's line (cf. p. 164). The Karlstad branch also demonstrated its radical lack of compromise when the handing in of notice was a fact (cf. p. 168). At the same time the policy conference was the most widely recruited assembly called to central discussions during the collective negotiations. The communist agitation and the radical views expressed at the policy conference were never to be sufficiently strong later to have any effect in the representative assemblies. The representative council met twice during the collective negotiations and, when the heat was on, followed the line selected by the negotiating team and the executive board. The agitation against this line had to be carried out in the branches of the union.

What was especially sensitive for the board was of course such opposition as threatened solidarity during the strike. It appears quite clearly that in certain quarters the communists carried on a campaign against the leadership and for a much harder line in the struggle against the employers.[98] Thus in January 1931 branches 36 and 114 in Karlstad sent an open letter to the branches of the union urging them to repudiate the leadership of the union and the negotiating team with the purpose of pushing a much harder line than the readiness to compromise of the leadership, which was certain to end in the reduction of wages by some percent. It is natural to interpret this as factor in the activities of RFO (Red Union Opposition) carried on from 1930.[99] This open rebellion was condemned in no uncertain terms by the leaders in a circular headed "A crime against solidarity"[100]:

> An 'open letter' has been sent out to the branches of the union to all textile workers, signed by branches 36 and 114, Karlstad. In this 'open letter' is first advanced the lying suggestion that the 'executive board's point of view before the breaking off of negotiations was approximately the status quo', and it is stated that 'one can easily understand that the compromise will come to be a reduction of quite a few per cent'. Following this a completely new proposal is presented as a basis of negotiation and finally the union's members are exhorted, behind the backs of a leadership appointed by Congress, to organize collective negotiations and to select a new negotiating team.

The leadership stated that the Karlstad branch was "in a triple sense guilty of a crime against solidarity". In addition they had also obviously violated the union's standing orders by setting up new directives for the negotiations. In a sharp conclusion the rebels were branded:

> Comrades! Guard our solidarity and show the door to these diversive elements who apparently are only driven by political party considerations. Don't let them destroy the struggle of the textile workers. The most fitting answer to the 'open letter' from both of the Karlstad branches is to return the paper with an exhortation to learn some manners.

In a leaflet entitled "Fight against the organizers of defeat", the Swedish communist party went to the textile workers in its own name and presented ("agreed with") the Karlstad program.[101] It was stressed that it was about a fighting program, which even the revolutionary textile workers in Stockholm accepted. Most of the demands put forward were standard demands from the communist party.[102] This applies among others to the demands for shorter working hours and the abolition of piecework. Certain demands appeared to lie somewhat closer to the actual working situation of the textile workers (demands for equal pay, for wage increases, for holidays, for leave when pregnant, etc) but even they were worded in a stereotyped manner without any adaptation to the current negotiation situation. A 25% increase in hourly wages had been demanded as the opening offer in the negotiations, but some way into 1931 it must have been quite clear to everybody that such a raise was a total impossibility. The communist move was a demonstration policy with but slight connection with the negotiation situation in which the Textile Workers Union was placed. The dissension therefore never created any real threat to the leadership of the union. The circular, which stigmatized the action of the Karlstad branch, was a way for the leaders to maintain and show their authority. Nevertheless it was not possible to disregard the arguments put forward by the communists. It was only with manifest difficulty that the leadership could defend the signing of the new agreement in spite of the members' opinions. Gratefully the communist activities were used as on argument in the explanation:

> The situation was further made more difficult because the ... communist newspaper NY DAG openly discussed what was in the circular and what was reported at the meeting about the union's finances. By this open treachery the employers were spurred to make things much more difficult than they would have been otherwise. It cannot be excluded that the settlement made across the board might have been better if 'Ny Dag' had not revealed the financial position of the union. 'Ny Dag's' treachery with regard to this belongs to one of the darkest chapters in the history of the Swedish trade union movement.[103]

In spite of their relative unimportance in the union the communists could be utilized for some of the blame for the settlement made by the union's negotiating team against the expressed will of the majority of the members.

8. Summary: Piece wage problems in the textile conflict

In the above the textile conflict has been subjected to an intensive study where every phase in the work of negotiation has been followed carefully. It is obviously reasonable to ask why such a meticulous examination was made.

The demands and the turns and twists of the textile conflict not only reflect the conflict of opinion at that time but the demands held back for a decade. This alone justifies a detailed explanation, especially as the conflict in contrast to many other struggles of the labour market during the 1930s have not been subjected to any great attention in earlier research. When after ten years of inactivity the textile workers took the offensive the most important demand was an increase in the scheduled wages, but even the position of the piece wage workers played a great part. The piece wage excess should be set at a higher level by means of a central agreement. In addition the workers' local influence at piece wage negotiations was to be strengthened by the acknowledgement that the union organization should play a role when the agreement was made. As further support for the weaker party must be regarded the determination of the grounds for the setting of piece wages, which was expressed in the demand for production investigation of a kind that was easy to understand. Here was not only to be stated what the grounds for the piece wage should be but also that the intricate time and motion studies, which already because of their complexity seemed to be something for the employers alone, were to be kept outside the activities of the textile plants, at least as a tool to be used in negotiations about pay.

The employers who of course in the opening stage of the negotiations were on the defensive, soon found that the economic position developed in their favour and got together on the main target of reducing wage costs in the first place by a reduction in the piece wage excess. The effect of this revision, to be carried out centrally, was to be safeguarded by a provision in the agreement making it possible for the main organization to prevent an undesired wage drift in piece wages. As a complement to this overall cutting of the wages paid to the piece wage workers the employers' organization was also interested in protecting local freedom of action for managements of plants. No central agreements about influence were to approach too closely to the principally unrestrained freedom of the piece wage agreements; a reasonable attitude from their side as such an arrangement maintained the advantage of the locally stronger party. Still less were the employers prepared to yield any part of the right of management to decide how the conditions of production should be arranged. Both of these reasons were sufficient to reject energetically the regulation that could be thought to threaten the rights to utilize such efficiency measures as time and motion studies.

The very careful account of the textile conflict's rounds of negotiations are also justified for pedagogic reasons. Certainly these exchanges of views during the early 30s show many of the arguments and disputes as well as the manifestations of desires and lines of action that were to dominate the collective negotiations during the following decades. There are to be discovered the workers' demand for piece wage compensation and there is emphasized also the employers' idea of provisional piece wage schedules, which were to give greater flexibility in the setting of wages, together with the model advanced by the conciliator for the revision of piece wages that was to play a great part in what was to come. The trial of strength demonstrated by the conflict also puts into relief the relations between the parties during the remainder of the period covered by the investigation. It was an ever present experience for the actors.

Furthermore the textile conflict in the winter of 1931 was also one of the few occasions when a more radical opposition in the Textile Workers Union managed to make itself heard. An this also shows in the central material on which this investigation is constructed. It was a radical, partly communist, opinion of members that urged termination of the agreement and established the demands for which the workers went on strike. Later the leadership gained an increasingly firmer hold on the helm of the organization, but even when the strains and stresses of the conflict had been felt for many months there was still a greater offensive spirit among the members than among the workers' leaders. But the latent strength of the latter was demonstrated when the negotiating team accepted the responsibility for a settlement, which to a great extent had been rejected by a predominant majority of the members.

Among the questions pushed by the radical majority is again found a sharper attitude to rationalization and time and motion studies than those that the reformist leadership held. As the development during the years immediately following was to show this sharper attitude did not provide the opposition with any chance to gain any greater influence. The executive board's stable and conditional social democratic acceptance of the forms of rationalization was to show itself to be more in accord with the ideas of the members.

When it is finally considered how the questions of the setting of piece wages and time and motion studies were utilized in the negotiations, it appears obvious that these factors were subordinated to the questions directly associated with wages and given priority by both parties. It was with regard to the wages paid that the main differences lay. Such a priority was obvious for the employers, who really only admitted questions of wages as negotiable, and this can easily be explained on the workers' part by reference to the instrumental idea which was strong in the Swedish trade union movement: the main function of work was to provide earnings that could enrich the workers's existence beyond the factory gates.

It was this clarity about the primary role of pay questions that lay behind the conciliator's attempt to create a difference between questions of earnings and questions concerning the general provisions of the agreement. The workers, especially, objected to this arrangement most strongly. The wanted a solution of all the disputes at one and the same time. This demonstrates that the demands for changes in the general provisions put forward also had the function of a collateral to be used for the purpose of compromise: they could not be detached from the framework of negotiations, a circumstance that in this connection also serves to explain the rather detailed account and analysis of the textile conflict.

CHAPTER 9

From 1931 to 1936

1. A time for cutting wages 1931–1933

The agreement concluded in March 1931 continued in force until 1936. The costly conflict and unfavourable trade conditions curbed the offensive spirit of the workers. In May 1932 when the agreement was once again on the agenda, the representative council of the Textile Workers Union decided it was not even worthwhile to initiate a discussion about demands or termination of the agreement.[1] In the summer of 1933 the council discussed a possible notice of termination but concluded that this was out of the question.[2] This opinion was shared in the autumn by the executive board.[3]

In the following year a somewhat more open discussion was carried on with the same result. The representative council recommended that notice should not be given.[4]

In 1935 this council again considered the possibilities of demanding changes in the agreement with the hope that preliminary negotiations could be carried on between the parties without the agreement being terminated. As the leaves of autumn stated to fall the representative council recommended that the union should start such negotiations and this occurred during September. At these the workers also advanced a demand for alterations in the piece wage paragraph. However, the employers refused to discuss any alterations in the general provisions unless notice was given. The representative council therefore authorized the executive board of the Textile Workers Union either to give notice to terminate the agreement or to try to get it extended.[5]

As was expected, nor was the Textile Employers Association in the summer of 1931 inclined to terminate it for a slight probability of obtaining new wage reductions. However, since the spring the Employers Confederation had discussed the possibility of joint action in a number of areas covered by agreements (see pp. 164 et seq.). Following an enquiry from the Employers Confederation in August the Textile Employers Association decided under certain conditions not to stand aside from a joint notice of termination of the agreements. In this case it was necessary that most of the large unions agreed and that settlements should be made across the whole field at one and the same time.[6]

Feelings in the Employers Confederation were obviously in favour of starting the big offensive. Therefore the general council of the Textile Employers Association proposed at an extraordinary meeting of the association that it should in loyalty to the main organization give notice to terminate the collective national agreement with the Textile Workers Union.[7]

On the following day the general assembly of the Employers Confederation met. Axel Bergengren reported at this the association's decision but also made a definite reservation that the association intended to try to reach a settlement by voluntary means if this was possible.[8]

Thus the Textile Employers Association only six months after the greatest conflict in the history of the indusry again started offensive negotiations. The preparations included a new proposal for the collective agreement with the Textile Workers Union that in its main points was characterized by the desire to regain the ground where it had been necessary to make concessions in the spring and to thrust further where goals had not been achieved. So the general council decided that the association should propose a cut of 10% in hourly wages and a revision of piece wages so that an average piece wage excess of 15% was obtained. The provision on production investigation was to be deleted.[9]

The Textile Employers Association was arming for a new struggle. But intervening events upset the plans. Two days after the council's meeting, on the evening of Sunday 27 September, the Swedish government decided to follow Great Britain's decision to leave the gold standard. The following day the National Bank of Sweden raised the bank rate to 8%. This drastically changed the conditions for the economic situation of the industry that had been assumed by the Employers' Confederation in its decision about an effort to reduce wages. On the 29th the Employers Confederation decided to release the varous associations from their undertakings to take part in a joint offensive. The way was thus clear for every association to reconsider its decision about termination of the agreements.[10]

The general council of the Textile Employers' Association met the day after. Opinions differed about the situation that had arisen. In order to clarify how the members of the council regarded the situation it was decided to take a test vote. This showed seven members in favour of giving notice to terminate the agreement and five agains. The majority thus wanted to continue with the planned offensive line, but the minority was so strong that it could not be ignored. In this position Erik Paues, who was also chairman of the Swedish Association of Clothing Manufacturers (Konfektions industriförbundet), had a decisive influence. With reference to the strength of the minority he recommended that notice should not be issued. This also became the decision of the general council, against which the chairman, the vice-chairman and a further member entered a reservation.[11]

The threat to industrial peace in the textile industry was thus avoided at the last minute by the narrowest margin. The great wage-cut offensive by the Employers Confederation was also considerably weakened by the actions of other associations, even though some of them continued along the path prepared. This effect of leaving the gold standard must of course be taken into consideration in assessing the social results of the government's economic policy.

Leaving the gold standard did not cause the fall in the rate of exchange expected by the Employers Confederation. Therefore in the following year, 1932, the idea of a greater joint action to obtain a decrease in the wage level of industry was again considered. So many as possible of the agreements that had run out at the end of 1932 ought to be terminated said the board of the Employers Confederation. When the plans were considered by the general council of the Textile Employers Association, Gustaf Söderlund, managing director of the Employers Confederation, was present and developed the confederation's view of the tactical and strategic position of the employers. The fall in the exchange rate had not occurred as expected following the currency reform of the previous year and this had contributed to the shelving of the great action for the reduction of wages. In order to stabilize the purchasing power of the Swedish crown a fall in prices and wages was now required, he maintained. But there were also other reasons suggesting that a more comprehensive notice of termination of agreements should be issued. It had appeared at negotiations during the year that the workers to a great degree accepted the necessity of wage reductions. If the Employers Confederation now omitted to take the offensive the workers could, thought Söderlund, get the impression that the employers were satisfied with the wage situation. This could in its turn cause the workers to initiate action for better wages and benefits at the slightest improvement in the economic situation. Therefore the employers should now take the offensive in a joint action. There were certainly circumstances suggesting that to give notice was unadvisable. A greater struggle could be expected on the labour market after the new year and the development of the currency situation was very uncertain. But these doubts did not affect Söderlund's final assessment.

Following this presentation the discussion was started by Pauer with a survey of the economic situation and future prospects for the textile industry. His speech ended with a weighty recommendation to give notice of termination:

> According to the managing director the association has large interests to safeguard by giving notice. If such a notice is not given now it seems to be uncertain when a better opportunity will arise.

During the course of the debate it was seen that a strong majority among the general council's members present favoured joining the front being formed by the Employers Confederation. With somewhat greater doubt was regarded the possibility that the Textile Employers Association would carry on an individual struggle.[12]

However, not even in this year did the preparations result in the wide action intented by the Employers Confederation. Presumably the negative attitude of the Metal Trades Employers Association contributed to this decision (cf. p. 81). In certain areas notice to terminate the agreements was given and especially in the building industry this was to lead to a lengthy conflict, but the textile industry was not to be involved.[13] Neither the employers nor the workers of the textile industry gave notice and the agreement was prolonged until the end of 1933.

Discussions were carried on again in the following year. Coordination with the Employers Confederation was the same, the managing director maintaining that a real reduction in wages was desirable (25 %) but that the feeling in the association was not sufficiently clear-cut for notice to be issued. At the same time he was extremely sorry that the association, influenced by the changes in the currency situation, had hesitated the grasp the opportunity of wage reductions in September 1931.

Following a careful discussion the general council decided in view of the unity existing in the association to try to get the agreement prolonged. If a large majority in the Employers Confederation was seen to demand a joint wage reduction action the matter could of course be reconsidered.[14]

2. The situation reversed

In 1934 the improvement in trade within the industry had again changed the conditions for negotiations between the two parties. It was now clear the employers had missed the chance of utilizing the depression over and above the bills that had been drawn on it in the late winter of 1931. None of the districts of the Textile Employers Association now asked for any notice to be issued. The general council also decided that nothing of this kind should be undertaken other than in the shape of a counternotice should the other side open the offensive. In preparation for this eventuality a proposal for a new agreement was approved.[15]

When the time arrived for the general council to consider its position on the agreement, whilst the meeting was in progress a delegation from the other side requested a discussion with the leaders of the Textile Employers Association. The session adjourned and a meeting with the delegation took place. The representatives of the Textile Workers Union reported that feeling in the union was in favour of notice being given to terminate the agreement. But this threat to the peaceful industrial situation could be avoided if the employers were prepared to meet the workers' demand concerning certain medical care benefits that looked like vanishing in conjunction with the changeover to the new national health arrangements.[16]

When the general council resumed its internal discussions it was decided to accede to the demand made by the Textile Workers Union.[17] Thus the threat of notice was avoided. The course of events highlights how the position of the employers had weakened since the previous year when Paues was still able to talk about wage reduction demands of 25% without losing credibility.

But how did the draft agreement look that the Textile Employers Association had decided to present if notice to terminate the agreement really had been given? Quite a few striking changes had been made in the provisions of the piece wage paragraph on the grounds for the calculation of piece wages. The target of the agreement, a *minimum* piece wage excess of 15%, had again become an *average* of 15%. With regard to the grounds for higher pay in certain cases it was stressed in the proposal that this would occur after being examined by the management, a change that had no equivalent in the current agreement and which marks more strongly the employers' unilateral right of decision. In addition an important change had been made in the grounds stated for such higher excess. The current agreement stated:

> ... in which case special workers, who are given more responsible jobs, either heavy such or with regard to more machine units or which demand special accuracy and skill, should receive a higher excess.

In the employers' proposal the text read as follows:

> ... however, the workers who are given more responsible work or work that demands special accuracy and skill, may be able after consideration by the management to receive a higher excess.

The change in the verb is noticeable: "should receive" becomes in the new text "may be able to receive". The grounds "heavy work" and "more machine units" are not mentioned any longer. In the present context the last mentioned is especially interesting. Rationalization in the textile industry during the interwar years mainly found its expression in the increased number of spindles or looms per worker.[18] If every such change automatically and contractually were to be followed by the right to a higher piece wage then naturally the financial saving and thus the incitement to rationalization methods become considerably less. The same line was to be followed by the recurring proposal to delete completely the point in the agreement dealing with production investigation.[19]

It is quite clear that in those extracts dealt with here it was the ambition of the employers to ensure an effective "paragraph 23", (cf. p. 15), to ensure their own control of the means of rationalization and its results and to prevent the piece wage excess that arose through rationalization measures from creating a larger gap between the wages of the piece workers and those who received an hourly rate of pay.

The year 1934 had been a turning point. The turn taken by trade lessened the employers' tendency to take the offensive and added fuel in a corresponding degree to the desire of the workers to advance their positions.

During 1935 the change in climate became more noticeable. As early as May the representative council of the Textile Workers Union discussed the agreement and it was clear that the demands and desires for changes were many: it was a case of a need restrained by the years of crisis. But it was not assumed that it was necessary to give notice. The technique of preliminary negotiations could also be utilized.[20] A variation of this theme had of course been utilized with a certain amount of success the previous year when the unions had refrained from giving notice in exchange for discussions on the problem of the new national health organization being started. These had in fact led to an agreement being concluded in June 1935.[21] So the experience was comparatively good.

At the beginning of September the representative council decided that the Textile Workers Union should try to arrange preliminary negotiations on a comparatively extensive list of demands.[22] By the following day a delegation from the union had agreed with their opposite numbers that negotiations should be initiated on 17 September.[23]

When the parties met, Gösta Wennström of the union started the ball rolling by referring to the good experience of preliminary negotiations that had been made within other unions. During the year such preliminaries had resulted in improvements for the workers. He thought this must be interpreted to mean that the employers admitted that better times were coming and that the workers had the right to their share of what the improved economic conditions brought with them. Wennström then went on to explain the workers' demands in 12 points, of which two especially attract attention in this context. One was the demand for piece wage compensation for those workers on hourly pay. How large the compensation was to be was not stated—it was an agreement in principle they were interested in here. The second point concerned application of the general provisions of the agreement about the piece wages in paragraph 7. Here the workers claimed agreement from the employers on (1) that all work in the main body of the textile trade should be put on piece wages—only when negotiations broke down should the employer have the right to have the work carried out on hourly pay; (2) in the increasingly appearing rationalization measures in production (for example the transfer from a smaller to a larger number of machines per worker) the piece wages should normally be terminated by notice so that negotiations could take place and (3) that the worker whilst awaiting the piece wage agreement should be paid his average wage instead of at the old piece wages or hourly wages.[24]

After Wennström had put forward his claims the employers pointed out that the delegates of the Textile Workers Union did not have the necessary authority to make an agreement on the spot and for this reason refused in all respects to state its position. In addition certain of the demands put forward were of such a principle character that they could not suitably be discussed at preliminary negotiations. Among these more far-reaching questions was included that concerning piece wage conditions. Following a discussion on the remaining points at issue without any decisions being made the meeting was adjourned until 26 September.[25]

In the Textile Workers Union the course of the preliminary negotiations was discussed by the representative council and the negotiating team appointed earlier received authority either to carry the negotiations to a conclusion with certain improvements achieved in the agreement or to give notice to terminate it.[26]

The general council of the Textile Employers Association met both before and immediately after the preliminaries with the workers' negotiating team. The working committee was given the task of preparing a draft agreement in the event that the other side should give notice but this was to be regarded purely as a preparatory move. Initially it was counted on continuing and bringing the preliminary negotiations to a conclusion and the general council confirmed the association's line: not to permit negotiations about questions of a more principle character but on the other hand not to advance any other than quite modest claims themselves, which however included that a settlement should be possible on the spot.[27]

Negotiations between the parties were resumed on 26 September when certain aspects of the piece wage question were discussed. With regard to the demand that current piece wage agreements should be terminated by notice, for example on transfer from fewer to more machine units, the employers pointed out that the parties had been in agreement before the Labour Court that such notice was unnecessary.[28] Since then the workers had altered their point of view and the employers declared they were now willing to go along with them on that clause and agree that a new agreement should be made that applied to the new conditions whilst the old could continue to apply in those cases where production returned to the older conditions.

The second demand taken up meant that the workers who awaited an agreement on a piece wage should not be compelled to accept hourly pay as soon as the local negotiations had broken down but only when negotiations between *the main organizations* had shown themselves to be abortive. During the waiting period the workers should receive their average earnings so that the income level was not permitted to decrease before the main organizations had had the opportunity of deciding on the matter. The employers opposed the proposal with reference to the fact that there was no support for it in the text or meaning of the agreement. The proposed procedure would lead to an unreasonable number of disputes being referred to the main organizations and these lacked the resources to deal with such a number. The spokesman for the Textile Employers Association suggested another procedure:

> The most suitable procedure should be that the workers—as has already occurred in a large number of cases—for a certain period, for example at least a month,

worked on trial at the piece wage rates, which the employer had proposed with the help of his time and motion studies. It was then up to each of the parties freely to call for negotiation in and for the adjustment of the piece wage proposals thus presented.[29]

A number of interesting aspects arise at this stage of the discussion. The idea of the workers' proposal was to try to protect the earnings-level of their own groups by demanding central negotiations before piece wage reductions actually occurred. It was reasonable to think that the greater apparatus associated with central negotiations was enough to restrain the employers from changes in piece wages that had a less sound basis. Behind this barrier was constructed a further line of defence consisting of the weight and knowledge of the main organization. Every real action for reduced pay could be taken up by the central organization. The influence of the union was to guarantee the retention of the earnings-level.

As an alternative to this extension of the central organization's influence the employers advanced the idea of a trial period at the price proposed by the management. The "correctness" of the piece wage, that is to say that it was within the limits laid down by the agreement, would be guaranteed as it was based on time and motion studies. So here the employers argued for the idea that time and motion studies were objective methods for the calculation of piece wages. The supposed scientific character was permitted to draw a veil over the basic conflict of interest.

The preliminary negotiations resulted in an agreement across the board in which the parties undertook not to give notice to terminate the agreement during the current year.[30] Additionally a number of changes were made in the agreement to the effect that certain of the workers demands were agreed, but not in any of those clauses in the piece wage provisions dealt with here. The preliminary negotiations were undoubtedly successful for the workers who obtained some of their demands and to a certain extent also for the employers who certainly were unsuccessful in obtaining their alterations but who in any case were able to avoid notice being given in an unsatisfactory trade situation.

3. Summary

With regard to the relations between the two parties in the textile sector the period between the spring of 1931 and the autumn of 1936 may be summarized in the following way.

After the sufferings of the great conflict no extensive negotiations on a new national agreement between the parties of the textile industry were undertaken during the first half of the 1930s. From the spring of 1931 until towards the end of 1933 the economic position was regarded as favourable to the employers. The management of the Employers Confederation also prepared itself for more extensive collective negotiations covering a number of areas of agreement. The Textile Employers Association allowed itself to become enrolled in these plans with a fluctuating enthusiasm. External factors or the doubts raised by important members, however, upset these plans for overall joint actions from the employers' camp.

Nevertheless, the planning alone shows that the sense of purpose within the Textile Employers Association, which would have been decisive for the demands put forward by the association in the event of negotiations. In the autumn of 1931 the employers were prepared once again to try to reduce nominal wages by cuts in both hourly wages and the target point of piece wages and to regain freedom in the calculation of piece wages by deleting the provision on production investigation from the agreement.

Prior to the negotiations that could be expected to occur in the autumn of 1934 the employers had even increased their list of demands. In addition to the old demands was added an attack on the specially high piece wage that the worker could obtain for tasks that were specially demanding.

During the following year trade conditions had markedly turned in favour of the workers and so the employers' reaction became completely defensive. It was now the other side that tried to advance its positions. The volume of piece wages was to be increased so that almost all work within the main body of the textile trade was put on piece wages. For the remaining workers on hourly pay an agreement in principle on compensation for piece wages should be made. One clear expression of the newly-won strength of the workers was the demand that piece wages really should be put under notice in such rationalization as required more work: the right of the workers to their share in the profits of rationalization was to be safeguarded. Nor was the employer, on the other hand, to be able to lower the level of earnings whilst awaiting new agreements on piece wages: in some cases the workers demanded that the average earnings should be paid. The reply made by the employers to this demand is worth noting: it proceeded from the assumption that new piece wages should be set with the assistance of job studies and offered a trial period for every new piece wage during which the synthetically calculated earnings of the piece wage could be compared with actual experience. The proposal implied the first suggestion of adapting the piece wage provisions of the national agreement to the terms of time and motion studies.

CHAPTER 10

Disputes about the Bedaux system 1933 and job studies 1947

1. The problem

The differences of opinion about job studies between the employers and employees of the textile industry, noted in the collective negotiations in the early 1930s were fully revealed by a dispute at Rydboholms AB. This originated in an attempt by the company in the summer of 1933 to introduce the Bedaux system and the dispute was settled by the decision of the Labour Court in the following April. The course of this dispute will here be subjected to a comparatively detailed description and analysis, where not least the Textile Workers Union's intentions and the role of the Labour Court can be shown.

The Bedaux system was a kind of rationalization program with ideas about operational control; the measurement of work output and pay were integrated.[1] The system was introduced to the company by a special firm, AB Nordisk Bedaux, and was also used in addition to Rydboholm at Malmö Strumpfabrik AB.[2] Even at the latter company attempts to introduce the system aroused opposition from the workers and a dispute developed there parallel with that of Rydboholm. But with the Malmö company the case was never carried to the Labour Court as the main organizations were agreed that the system in its entirety was unsuitable.[3] That part of the Bedaux system on which the workers concentrated their attacks in the Rydboholm dispute, and which the employers made a point of defending, was precisely job studies. These studies constituted an important part of the Bedaux system, but not all of it, and at the same time were also an important technique in rationalization in other systems than the Bedaux system.[4]

2. Sources of a dispute

In July 1933 Rydboholms AB initiated a rationalization of the company. A new organization was created and a different system was introduced for the calculation of workers' wages. So far as it is possible to discover no information was given to the personnel about the work to be carried out until the beginning of August. It is quite certain that by this time the opposition of the workers to the plans was already aroused. The management now appealed for trust and collaboration between the parties and presented to the workers what it was intended to do. In this presentation it was pointed out that the increasingly greater price competition had made rationalization necessary. The management emphasized that the job studies, which were a condition for rationalization, did not have the purpose of supplying reasons for cuts in the workers' wages. Instead, in some cases, they might possibly bring about increases, especially for workers on hourly wages. Thus were the ideas about questions of wage-setting presented and at the same time it became clear that the Bedaux system was to be introduced even though this was not said plainly:

> The reorganization means a systematic study of all work at the company's plants. On the basis of this study the work will so be reorganized that the working time with calculated time for recovery will be utilized. The normal work output per hour is described by 60 units and is paid at the normal hourly wage. Payment for a higher output of work than the normal goes up to 75% to the worker himself.[5]

It is quite clear that here the company went to work in a somewhat unfortunate manner. The information and urge to confident collaboration first came when job studies had already been started at the plants and the negative reaction was certainly an established fact. In addition, in almost cryptic terms, was proposed a reorganization on the basis of job studies and the calculated "recovery time" so that working time would be better utilized. Implicit here was the insinuation that working time had earlier not been used quite properly. The reorganization was to affect the grounds for wage setting, it appeared, which of course could also be expected to arouse suspicion, especially as the workers had already in the national agreement been awarded the right to influence and control over the usual manner of carrying out production investigations through production statistics prepared later.

For the changeover to the Bedaux system the company had employed experts led by N. E. Persson, an engineer, who represented the consultants AB Nordisk Bedaux. In order to carry out his task Persson gave instructions that certain statistical information should be provided continually by the workers whilst the work was proceeding. But they refused to supply the information requested to the job study men. Thus the dispute was established and the parties, the company and the local branch of the Textile Workers Union, branch 65, contacted their main organizations.[6] How defective the information was that the branch had at that time appears indirectly from a turn of pharse in the minutes of the Textile Workers Union. There the situation is described as "A work of reorganization already started" which, among other things, refers to the introduction of a new pay system, *presumably the so-called Bedaux system.*"[7] (Author's italics).

The reaction of the Textile Employers Association is not known. On behalf of the Textile Workers Union, its representative had informed the Rydboholm branch that the system in question was contrary to the agreement and that the workers should not approve it. This attitude was also approved by the executive board of the union.[8] So the workers at Rydboholm clearly had the support of their organization in opposing the new reorganization. There is reason to presume that this unhesitating support from the Textile Workers Union was a necessary condition for the dispute to be carried further. The question that arises here is whether the union was not trying, assisted by its interpretation of section 3 c of paragraph 7 in the agreement, to diqualify once and for all the use of job studies as a basis for the setting of piece wages in the textile industry. The unrest caused by the Bedaux system at Rydboholm presented a suitable test case.

The main organizations of both sides were now brought into the dispute and central negotiations were prepared. Prior to these the Textile Employers Association obtained supplementary information from the company. In the company's answer interesting interpretations of what the dispute was about and how it had arisen were supplied:

> ... In the course of conversations we have pointed out to the chairman of the local trade union the absurdity of the workers refusing to supply the required information about production to the management and have declared that this concerns an organizational question, not a pay question, as we have never anticipated any new pay system without previous negotiations. The chairman reported, however, that the workers refused to supply the information on the orders of the Textile Workers Union and they could not relinquish their attitude without new instructions from the union's leaders. We presume that the Textile Workers Union's appointed representative has misunderstood the situation and hope that you will be able to succeed in persuading him to change his view ...[9]

In order to achieve the last aim Erik Paues went to work and contacted the Textile Workers Union by letter and telephone. At this point the line of the Textile Employers Association already begins to appear clearly. After having noted the company's attempts to rationalize through the Bedaux system, Paues then broke down the problem:

> You probably will already be aware of the significance of the Bedaux system. For ourselves we have come to the conclusion that it essentially consists of job studies that aim at increasing production by a simplification or rationalization of operations. Thus far from our side, of course, there are no objections to the system and so far as we are able to understand there should not be from your side any objection to the system in this part. The further construction of the system may naturally presume collaboration from both the workers and the organizations to that degree it concerns permanent conditions of agreement. In such circumstances it is obvious that Rydboholms AB has had no intention of denying that the workers should exert their joint determination on the laying down of working conditions, even those applicable in the case of the Bedaux system.[10]

Paues thus resisted strongly when it was a case of safeguarding the right to job studies, especially with regard to method development. But at the same time he also stressed the workers' justifiable demand for collaboration and joint determination in questions concerning contractual conditions and wage construction. It was not a case of defending the company or the Bedaux system in its entirety at any price, but the principle of job studies and the rights of management laid down in paragraph 23 that were the most important. Therefore Paues proceeded with a concession. The lists of production results that the company had started to display in the plants and in which the individual performances of the workers could be seen had made their blood boil. The Textile Employers Association had therefore urged the company to stop these notices.

From this concession Paues then tried to tempt with a carrot. The Rydboholm plants, he said, comprised dye works, a printing shop and preparation plants. For a long time the union had advanced a demand that workers in these production phases should also be able to receive piece wages. Perhaps said Paues, without giving any definite promise, the job studies in progress could result in piece wages even for the categories of workers concerned here. Paues concluded his letter with the request that the union's representative should urge the workers at Rydboholm to provide the information required by the management of the plant until further notice and until, should this be necessary, it was possible to conduct central negotiations on the matter.

In its reply the Textile Workers Union indirectly rejected Paues' splitting of the problem. It was considered that the job studies undoubtedly still had the purpose of introducing a system that in its parts and in its entirety was impossible to accept. Therefore the union was not prepared, at least before the matter was properly investigated, to compel the workers at Rydboholm to relinquish their attitude of not providing the information. The Textile Workers Union found further support in the argument that it was not the company but the consultants Nordisk Bedaux who were carrying out the investigations so poorly regarded. The workers had no obligations to or agreement with this firm. The union was, however, prepared to enter into negotiations with the Textile Employers Association.[11]

A week later the appointed representative reported to the executive board of the Textile Workers Union about what had been said in the exchange of correspondence and that the organizations had agreed to start negotiations on 25 October. The report also provided the board with cause for general condemnation of the Bedaux system. The union also sent out a circular to the branches warning them against accepting the system.[12] Thus even before the central negotiations had started the Textile Workers Union had carried the battle against the Bedaux system and job studies to a national level.

3. Central negotiations

The opposing parties met at the offices of the Rydboholm company on 25 October 1933. On the employers' side were the plant management, the rationalization consultants and the Textile Employers Association, represented by Paues, on the other representatives of the textile union's branch 65 and the Textile Workers Union under the leadership of K. J. Olsson.

The representatives of the management opened the negotiations by expressing the significance of the Bedaux system and the aims of the job studies initiated. It was an attempt to obtain what could be considered an average performance:

> ... what could be reasonably required of an average worker at a completely normal working pace and it was thus not the meaning here to require any forced work. The purpose of these investigations was not only to obtain a basis for the calculation of the earnings of the workers concerned buι also to make clear to the management of the plant whether there were defects in the organization of the work in the shape of lost time and such like.[13]

Thus there was no doubt about the purpose: job studies for the development of methods and the setting of pay. This clearly opposed the line put forward earlier by Paues. He therefore interrupted the negotiations and pointed out that these two things should be kept apart in the Bedaux system, "namely *partly* job studies, *partly also* rules for the calculation of pay". Only the first of these two things should be discussed at the present negotiations, the pay problem could wait until one of the parties later wished to have a special examination of the question. Job studies alone were a wholly necessary instrument for the employers who wished to remain competitive by means of modernization. As Paues saw the matter the workers could have no reason for opposing such an attempt.

Again the workers stubbornly refused to admit any justification in the way Paues split the question. So long as job studies in any way aimed at setting piece wages the workers were unable to accept them. If on the other hand the employers wished to utilize such a technique in order to discover defficiencies in their own organization, they were welcome to do so. But as a basis for the setting of piece wages it was still only possible for the actual production statistics to be accepted. This was what was meant, said Olsson, in what was laid down in the agreement about production investigation. The text of the agreement confirmed that the workers should have the right of joint determination of piece wage prices and have the possibility of checking the foundation on which the price offered was based. Neither of these demands was fulfilled by the Bedaux system, where the employers had completely locked the earnings and where the piece wages were based on calculations so intricate that the workers could not be expected to understand them. There was also the fact that the job studies were not carried out by the company's own employees but by outsiders.

Here the workers' side was extending its arguments. Olsson took his stand on the agreeement and thought he could show that the Bedaux system was contrary to the agreement. The only correct basis for piece wages was the traditional investigation of the production volume in a stated period. All other calculations of performance were rejected. It is also interesting to note the twist given to the argument about outside consultants by the link with "intricate systems"—"external specialists". What came to light here was a sceptical attitude to the current professionalization of the rationalization movement and a fear that outside experts would set the tone entirely. It was possible to deal with the white-collar workers of the company but the technical rationalization of the experts, who were masters of efficiency methods but had no experience of working conditions specific for the branch, would be more difficult to handle.

In the continued exchange of arguments the management of the company defended its line including the legitimacy of the Bedaux system's calculations on pay. But that the management here was about to collide with a representative of its own organization became quite clear. Paues declared quite candidly that the technical sections of the Bedaux system about pay were contrary to the current textile agreement; thus far the employers' and the workers' organizations were in agreement. Paues thought that on the other hand the company must be entitled to continue with job studies and the workers were responsible for providing the information requested. Finally Paues suggested that in future the workers should not raise obstacles to the work of modernization carried out by the company. Later the union was to deliver a written reply to this. The meeting then closed.

What lay behind Paues' nonacceptance of the Bedaux system was thus not the fact that job studies were used but rather the way wages were calculated. Apparently Paues and the Textile Employers Association preferred clear cut kinds of pay: hourly pay or straight piece wages.[14]

During the period that followed it was mainly up to the workers to take the next step. After a number of exchanges during which the local branch of the union in Rydboholm conferred with its main organization, the branch announced a decision.[15]

In a letter to the management of the company at the end of November 1933 the executive board of the local union stated that it was prepared until further notice to supply such information as had already been supplied to the company before the consultants' services had been utilized. The letter went on to say that the workers wanted nothing to do with job studies connected with the Bedaux system.[16] Refusal to collaborate and the blockade of information thus continued but now more directed towards the new factor represented by the consultants, Nordisk Bedaux. Information given earlier would continue to be given by the workers. No suggestion that the workers could assist in job studies was made.

Rydboholms AB replied with an acrid communication. That the workers gave way with regard to the customary information was noted. When they then declared that they refused all collaboration in the job studies connected with the Bedaux system, the management stated this was of no interest whatever: the workers had not even observed the offer made to inform them of the technique utilized. There were, continued the managing director, two parts of the Bedaux system that affected the workers concerned: the pay system and operational control. With regard to the former the main organizations were to be brought in so that it was not dealt with further. But in the case of operational control the company really had the intention to introduce such supervision: here there was no room for negotiation. If the trade union persuaded the workers to refuse the information necessary for this control, they then broke the agreement by interfering in the rights of the employers. The position of the consultants to the company and the union was finally nothing to do with the case. Persson, the engineer, had the management's full authority and confidence. The communication concluded with a request to the union to cancel the blockade of information. To this a written categorical answer was requested.[17]

But the categorical answer did not arrive. Instead a lengthy communication dated 3 July 1933 from the chairman of the union branch, Nils Erikson, justified the stand taken by the workers. Vigorously the change of having interfered with the rights of the employers was rejected:

> We have not disputed your right to carry on job studies or so-called operational control. The management of the plant has the right to employ as many engineers as it likes and these may make whatever notes they wish. *We have never tried to prevent this*. The only thing we have refused is to collaborate in the job studies ourselves. We hold the opinion that it is the employer's own affair to carry on such job studies. We have been unable to discover any contractual responsibility to participate in them. Our agreement applies to factory work and what belongs to it, that is to say not the work of foremen, engineers, office personnel, etc.

If the employer had transferred time studies to a firm of consultants it was therefore up to that firm to get the necessary information, thought Nils Erikson. In his conclusion he emphasized that the workers held to their earlier view of meeting the employers halfway and accepting negotiations about a simpler job study system and to start again to provide certain information.[18]

The position was unchanged but the tone of the written exchanges more acrid at the start of the new year 1934. It was again time for the central negotiations on the question of whether it be possible to compel the workers to collaborate in job studies? Could the management be compelled to give up the Bedaux system under pressure not only from the workers' opposition but also against the opposition of its own association?

4. Continued negotiations. The Labour Court's decision

The new central negotiations got under way on 16 January 1934 in Stockholm.[19] When the correspondence exchanged between the parties since the negotiations in October had been dealt with the employers presented their case. Now much more weight was given than earlier to the employers' right to direct and allocate the work.

The employers maintained that because of this right the workers were responsible for providing the information with which the management was concerned. Following this the argument on the situation regarding the consultants was rejected as being untenable. Then the differences between the pay questions and technical operation investigations, including job studies, were emphasized anew. Finally the employers presented a memorandum on the information required from the workers in the different departments of the factory.

At this stage the arguments of the employers were wholly flavoured by the line Paues was trying to push, which stressed a sharp difference between technical questions and those where interests were involved. The basis of the argument was "paragraph 23". This provided substance and meaning to this separation of the two questions with a clear admission of the right of negotiation questions of pay but with an equally clear maintenance of the exclusive right of decision of the management in questions relating to production. In the line pursued by the management, that is on a "lower" organizational level but nearer to the actual processing, the workers and the local union, this dividing line was not quite so clear. In the communication of 9 December from the management an admission was made unreservedly on the intimate relation between production and pay questions at the same time as the workers in a patriarchal manner were invited to take an interest in job study methods.

On their part the Textile Workers Union—now represented by Gösta Wennström, who was just taking up the position of acting union representative, and the representative Ernst Rosenlund—referred to the argument presented earlier. The workers had no obligations to Nordisk Bedaux, the division of the question was rejected—with regard to the company's arguments, on good grounds it may be thought—and it was stressed that the workers did not oppose job studies as such but considered they had no reason to assist in them. An argument now introduced was that the amount of information required took time from production and thus negatively affected the workers' earnings on piece wages. In reply the management said that to give this information only took an insignificant amount of time and that it was up to the workers to give notice of termination of the current piece wage agreements if they thought that earnings-potential had changed.

The central negotiations did not result in a solution of the dispute which now completely concerned the right of the management to compel the workers to collaboration in job studies by virtue of paragraph 23. Both parties now equipped themselves for a continued battle before the Labour Court. The Textile Workers Union called in their former representative, K. J. Olsson, who was on leave. Assisting him he had August Magnusson, from Scania, who had gained his experience in negotiations about the Bedaux system at Malmö Strumpfabrik. Olsson was to prepare a reply should the employers bring the union before the Labour Court. In addition he was given a free hand to take steps to obtain a favourable conclusion of the conflict. The union did not worry about the costs. It was also decided to try to obtain from the Rydboholm branch such examples as were favourable to the workers and which could be put forward in the Labour Court.[20]

The Textile Employers Association also requested further information from Rydboholms Aktiebolag.[21] At the end of February the employers were all set to proceed against their opponents before the Labour Court. In their application for a summons the employers gave an account of the dispute and at the same time developed their arguments in the well-known manner. No directly new arguments had been added unless one includes the reason for introducing job studies, which lay in the suspicion that "the working pace at the plant, at least at times, had left a great deal to be desired in some quarters". The splitting of the question was carried out with some degree of finesse. The Textile Employers Association stated that job studies aimed at organizational improvements often occurred with industry and that Rydboholms AB had decided to use Nordisk Bedaux for these and that the last named firm

in addition to the technical investigations on production likewise usually suggests to the industrial company concerned certain technical pay arrangements. The results of the investigations named, however, may be used largely independent of the pay system in force so that the question of kind of pay in this case is of secondary interest.

The application for summons debouched into the claim

that the Labour Court shall declare that the workers are responsible for providing the management of the factory with all of the informations about their work, which the management of the factory requires and in the way the management of the factory requires, and that the Labour Court shall order the Swedish Textile Workers Union and its branch No. 65 to instruct its members to comply with the regulations concerned of the company's management.[22]

In its reply to the Labour Court, when the Textile Workers Union submitted that the claim should be dismissed, the workers' argument also followed well-known lines: it was not the job of the workers to carry out job studies and in any case they had no obligations with regard to Nordisk Bedaux. The union no longer had any reason to reply to Paues' splitting of the question, said the union, as during the intervening period the company at Rydboholm had agreed not to introduce the system. Instead, some new arguments were presented by the union. One of these was completely bound up with the actual case. The Textile Workers Union alleged that it had been misled by the other party. The Textile Employers' Association had given them to understand in the middle of November that if the union could persuade the workers to start providing the customary information again then the matter would be settled. The workers had made the requested concession but despite this the management and the employers' organization carried the matter further.

The second new argument was of a more principal character. The Textile Workers Union maintained that the whole dispute was an attempt to compel the workers to collaborate in job studies in order thereby to destroy the basis of the fixed piece wage schedules and force the workers to terminate these with in many cases obvious consequences in the current trade situation.[23]

Rydboholms Aktiebolag was given the opportunity by its organization to comment on the defence put forward by the workers. In five points the company sharply rejected the argument put forward. The company denied that the matter concerned collaboration in job studies—it concerned the provision of information about operational control. The company opposed the idea that the workers in this were given any control function that could be considered contrary to practice. The company disputed the idea that it had ever been a question of introducing any new pay system. It stressed that there was no principal difference between the information requested from them in connection with the rationalization measures. The company stressed that the information requested was for the management of the factory and not for AB Nordisk Bedaux.

The first of these criticisms made by the company had undoubtedly some justice in fact, but the struggle on the principle level was interpreted by the organizations involved as a struggle about job studies. The suggestion of the company that it had never been its intention to carry out some kind of new pay system is obviously incorrect.[24]

The Labour Court had summoned the parties to a hearing on 17 March. During the hearing the representatives of both organizations advanced no new arguments for their points of view. A good two weeks later the court pronounced its judgment, which was in favour of the employers on all points. The management of the company had the right to a reasonable degree to request the information from the workers concerning the amount of work carried out that could be considered necessary.[25]

The jugment was not unanimous. The two members of the court, the representatives of the workers, were substantially in agreement with the majority but also wished to order the employers to make a special payment to the workers for the time it took to provide the information requested.[26]

5. Summary

There is much to suggest that the Bedaux dispute was selected as a test case by the workers in order to test the legal possibilities of carrying out the line the union had selected in the job study question: to behave as though job studies did not exist.

At quite an early stage the employers had understood quite clearly that the nub of the negotiations was the question of whether job studies were to obtain entry into the textile industry or not. Thus the Textile Employers Association's managing director was also compelled to a certain extent to repudiate the management of the member company by breaking down the dispute into negotiable and non-negotiable questions. It was obvious that in this conflict, as in other collective negotiations during the period, the main thing for the employers was to safeguard the freedom of action of company managements, to safeguard the right of management in accordance with paragraph 23.

In connection with the account of the collective negotiations in 1931 it appeared quite clearly that one reason for the workers' rejection of job studies was that these represented a certain system that was a synthetic way of calculating piece wage earnings and difficult to survey. For the workers it was important that the calculation of piece wages followed simple patterns that the worker without special resources at his disposal at the workplace could also follow. This motive was reflected in the Rydboholm dispute in the arguments directed especially against the presence in the factory of the consultants, AB Nordisk Bedaux. This constituted the interference of experts and could only favour the party whose commission the experts were carrying out.

Also in another way it was obvious how the workers regarded job studies as a threat to the existing balance between the employers and employees. The purpose of job studies, maintained the representatives of the workers, was in times of depression to compel such terminations of piece wage agreements as de facto prevented the use of the fixed piece wage schedules. If this was the case they would provide a way for the employers to attain the freer setting of piece wages they had been unsuccessful in obtaining through provisional schedules in the collective negotiations of 1931. To the degree the Rydboholm dispute really was a test case for the Textile Workers Union the decision was unambiguous: the Labour Court unanimously approved the principle of using job studies also in the textile industry. The court confirmed the workers' responsibility to assist in the studies although there was a certain degree of disagreement on how such assistance should be recompensed.

No real disputes about whether job studies were to exist or not hardly occurred after 1934 in the textile industry. In the annual report of the Textile Workers Union, which lists all such disputes as are made the subject of negotiations between the main organizations, one case is given from 1935 where the workers adopted a go-slow policy on job studies. This had resulted in certain workers being sacked. As a reprisal the local union had put up placards aimed at people applying for jobs. These were urged to contact the union. The dispute, which was carried to the Labour Court, did not really concern job studies or go-slow but the right of the local branch to put up such notices.[27]

But the same annual report shows how an increasing number of disputes from 1935 and the following years were solved by assignment to production investigation, production control or renewed production control. This factor suggests that some degree of unanimity was growing up about what could constitute an objective or, perhaps better, an intersubjective basis for the setting of piece wages. That such production checks gradually and increasingly also took the shape of job studies seems to be very possible.

Disputes about job studies were thus almost nonexistent at the central level.[28] One of the few following the Rydboholm dispute occurred at the end of the period investigated here at Malmö Yllefabriks AB (MYA).[29] For the latter day observer this serves as a contrast to the Rydboholm dispute and confirmation that the opinion expressed by the managing director of the Textile Employers Association in his letter to the Employers Confederation in 1946 was correct (see pp. 244 et seq.).

The situation leading up to the conflict was an occurrence at a weekend course about industrial democracy held at the beginning of November 1947 in Malmö. During discussions around the theme of job satisfaction the head of MYA, Th. C. Bergh, stated that job studies were then carried on in the textile industry and that they did not cause any particular irritation. The chairman of the union factory club at MYA, Bror Hansson, had opposed this. Job studies, he said, caused great irritation among the workers and the job study bureaux functioned as spy centres in order to inform management how the workers behaved. Another chairman of a union factory club, not named in the minutes, regretted the audibly wretched conditions at MYA and said that he had other not at all so negative experiences of job studies.

The exchanges between Bergh and Hansson caused the former to institute an investigation at the MYA factory in order to discover what could lie behind Hansson's ideas. Nothing appeared from this investigation that could substantiate Hansson's criticism and when he was asked he was unable to confirm his statement other than in general terms. An invitation to a discussion on job studies in the columns of the company's house organ had also brought no reply. It had caused the company to request negotiations in order to clear up the situation; Hansson's statement was considered to have damaged the company to a certain extent.

At the central collective negotiations instituted the preliminary investigator for the union, Bo Tillman, regretted that Hansson's speech, which he had read in manuscript, had been misunderstood as referring to job studies at MYA specially. He also admitted that neither The Textile Workers Union nor its Malmö office had received any serious complaints about the job studies. If the conditions at MYA were considered specially it could be pointed out that the workers had the idea that the tasks of the job study bureau and the management were intertwined in a way that served to conceal the picture of the job study bureau as "an impartial investigative organ".

A discussion then developed about the conditions existing between the technical personnel and the workers at MYA and about the question of to what degree job studies could be kept separate from other tasks of the management in planning production. A few incidents of a rather trifling character were mentioned as examples of a lack of confidence between the supervisory personnel and the workers. It was as a lesser part of this problem, said Bror Hansson, that he had touched on job studies.

The negotiations closed with a joint statement in a clearly positive spirit in which both parties rallied to the support of job studies:

1. The parties established that the audience at the weekend course must have understood Mr. Hansson's statement about job studies as referring to MYA. The workers' side said they understood that the management had reasonable cause to take this statement as extremely serious. In view of this it was also understood that the employers had called for an investigation.
2. . . .
3. During the main negotiations the parties had found that complaints against the fundamental arrangement and results of the job study activities at MYA had not occurred.
4. . . .

The workers' side said they understood that what was important was really the spirit in which the job studies were carried out. In view of this it was considered that a large apparatus should not be set in motion in order to investigate something belonging to the past.
In connection with this the employers added that it was always desired to encourage a good spirit around job studies.

The negotiations in Malmö at the end of 1947 removed all doubts. At the central level both parties had completely accepted that job studies were a part of normal activities even within the textile industry. A certain amount of difference and apprehension can perhaps be traced between the representatives of the workers at the central union level and some of those who were active at the local level. It is expecially interesting to note the high degree to which the representative, Bo Tillman, who in 1951 was to follow Gösta Wennström as chairman of the union, dealt with the ideas behind job studies: the job study office should really be the impartial organ for investigation, at the service of both parties when there were differences of opinion. The textile workers had in practice given up the old line of acting as though job studies did not exist.

CHAPTER 11

Collective negotiations 1936–1938

1. The build up and demands

Early in the spring of 1936 it was already obvious to the Textile Employers Association that the workers would give notice to terminate the agreement in the following autumn and so the association started to make certain preparations.[1] These continued during September when in the usual way the local branch meetings of the association were held to provide information about the situation and to sound out feelings among the members. The "technical conference" which was called for the middle of September may be seen as an expression of the gradually increasing complexity in the relation between production conditions and agreement questions.[2] Such discussions, which were markedly based on the technical conditions of production and which brought in participants at the senior engineer level had not occurred before. The technical conferences were to become a regular feature of the association's activities from 1936.[3]

The main task of the technical conference in 1936 was to deal with the problems that it might be expected the workers would raise in their demands, should notice be given. As a counterproposal to a general compensation for piece wages, the claim for which could be expected, the conference proposed that the association should put forward the idea of a grade structure for hourly paid workers where initially those who had a more expert task, to a certain extent supervisory, should be considered for an additional 10%. The conference put its argument into a more concrete form with a proposal for similar grade structures for the different industries.[4]

In this respect the proposal of the conference had obvious limitations. It was the case of a counterproposal for tactical reasons and dealt only with piece wage compensation for workers on hourly pay. Nevertheless it may be reasonable in these attempts at job evaluation to sense portents of the wage-ladder constructions that during the later 1940s were to form a noticeable feature in the design of the textile agreement. It is then worthwhile to observe the function of the technical conference of the employers' association as a generator of new ideas within the framework of the national agreement.

The general position, initially assessment of the trade situation, did not encourage the leadership of the Textile Employers Association to an open struggle. The general council decided not to give notice on its own initiative nor even to issue a counternotice should this possibility arise. Instead, a notice issued by the workers should be met with arguments about the positive development of real wages and the higher wages in the Swedish textile industry compared with those abroad. In addition the association should reserve the right to advance certain proposals of its own at the approaching negotiations. A few proposals for alterations were also presented to the council.[5] The decision not even to deliver a counternotice must be interpreted as an indication of the employers' intention to discuss only smaller changes within the framework of the current agreement.

As could be expected the carefulness of the employers was met by the offensive spirit of the workers. At the meeting of the representative council at the middle of the summer of 1936 all delegates agreed that the Textile Workers Union should give notice to terminate the agreement.[6] For various formal and practical reasons no decision was made before the ordinary Congress of the union that was held during the first half of September. Congress decided on organizational changes in the union, the representative council was wound up and replaced by a general council: a change mainly in name.[7] The general council, which held its inaugural meeting immediately, agreed with the recommendations about the agreement from both the representative council and Congress.[8] On the basis of the expressions of opinion the executive board decided for the first time since the conflict of 1931 to give notice of termination of the national agreement.[9] The new draft agreement, which the Textile Workers Union enclosed with its notice, contained changes both in pay schedules and the general provisions.[10] It was proposed that hourly wages should be increased by about 15%. The demand for piece wage compensation for hourly paid workers was again put forward and it was proposed this should amount to a 15% increase on the hourly pay. As the target in the setting of piece wages was given an excess of on average 30%.

The preamble to the piece wage paragraph was provided with a phraseology that more compulsively confirmed that the work so far as was possible should be put on piece wages. In so doing the period of validity should normally, unless otherwise expressly decided, be the same as that in the current national agreement, which meant that piece wages should be regarded as fixed and be approved by the main organizations, and also carried the right to require changes in the setting of piece wages in conjunction with notice of termination of the general agreement.

The rationalization paragraph was also made more stringent in the proposed text. Only in *important* changes in the conditions of a job should it be possible to terminate the piece wage agreements. This terminology was intended to deal with the two kinds of behaviour that the unions considered to be a misuse of the current provisions. On the one hand there were examples, it was maintained, to show that the employers carried out important changes without giving notice to terminate the piece wage, for example by increasing the number of machines per worker. On the other, there were also, especially during the depression when the pressure on wages was downwards, examples of how the employers even then made smaller alterations the excuse for new piece wage negotiations. When such notice was given the textile workers considered the old piece wage should continue to apply until a new agreement was made. It should be impossible for an employer by giving notice of termination of a piece wage to compel a worker to take hourly pay whilst awaiting central negotiations. Likewise, a guarantee was proposed in the new setting of piece wages; individual average earnings should apply until agreement had been reached.

The clause on the grounds for the setting of piece wages started off with a very categorical provision that piece wages *should* be based on the average production actually achieved. Such a construction should definitely banish job studies from the setting of piece wages. These should then be based on the traditional statistical production investigations, not on the prognostic methods of job studies. In the clause dealing with influence on production investigations, the expression "the workers' side" had been exchanged for the unambiguous "trade union concerned".

The important lines in the demands of the workers about the piece wage provisions may be summarized in the following manner. Piece wage work was to be introduced wherever this was possible because it gave higher earnings, on average 30% above hourly pay. The groups of workers unable to obtain piece wages should be compensated in part by wage increases. The setting of piece wages should be made so stable as possible, also including the increased influence of the trade union organizations. Job studies based on a more advanced theoretical method, which could be utilized for assessing individual performance in detail and which aimed at a prognosis difficult to control, were denied as a basis for the setting of piece wages. Instead, as the basis for the calculation of piece wages, the traditional production statistics were safeguarded, that is to say the results actually achieved. The combination of the different demands made by the workers was to carry with it a much more rigid system of wage setting, where the right to piece wages calculated in the traditional manner as well as the trade union's influence on both calculations and piece wage agreements had been laid down by agreement.

It is obvious that the employers reacted sharply against the constriction which they realized was threatened by the workers' proposal. The employers must retain the right to decide whether a job should be carried out at piece wages or hourly pay, said Erik Paues. This decision was dependent on whether the setting of a piece wage could be considered to lead to an increase in production or not. Nor could the piece wage be locked to the period of validity of the agreement. The attack on the rationalization paragraph as on the provisions about the basis of piece wage calculations must be rejected. It was necessary to maintain the employers' freedom of action and to get job studies recognized. A general revision of piece wages should be carried out when a reassessment could be made of what work was to be put on piece wages. The limit of 15% for the piece wage excess should be retained whilst the piece wage compensation, which apparently appeared to be inevitable, should be limited to the lower value of 10%. In this manner the increased earnings, incitement to the higher working pace of piece work, should remain.[11]

2. Negotiations under an imperial chairman

From the beginning the parties agreed to pursue negotiations under a neutral chairman and proposed Olle Ekblom, a conciliator who had many years experience of the textile industry, not least as a member of the conciliation commission during the conflict of 1931. Ekblom summoned the parties to opening negotiations at the beginning of November 1936.[12]

Prior to the negotiations it was necessary for the employers to transform their reactions to the proposals of the workers into their own counter-proposals.[13]

The leadership decided to stand firm on the main lines of the current agreement, including the stated hourly wages. The idea was that it should be proposed that the piece wage excess should be lowered to 10%. Apparently this was a case of an offer made for tactical reasons aiming at a compromise on the 15% limit of the agreement.[14] The employers also wanted release from the binding effect contained in the fixed piece wage schedules and to introduce schedules with a period of validity decided each time and which could be terminated by one week's notice. Further, the employers should request that the provisions on production investigation contained in the current agreement should be deleted.[15] On this point also tactical considerations may have played a part. The demand may be regarded as a tactical adjustment to the workers' further proposal to have the statistical production investigations approved as the only legitimate basis for the setting of piece wages. The introductory round of negotiations between the two parties was taken up by a debate in principle about the proposals put forward from each side. There was no real discussion that brought the parties closer but the negotiations were adjourned until later in the month.[16] The impressions of the workers' negotiating team about the employers' attitude are worth noting:

With regard to the general provisions, and especially in this case the piece wage paragraph, the employers' proposal contained demands for far-reaching changes for the worse from the worker's point of view. During the discussion it had appeared that the employers' side had a special partiality for holding forth on how important it was for managements to possess the greatest possible freedom of action, this with regard to the arranging of ordinary working hours and when it applies to the possibilities of carrying out shift work, and to adapt the piece wage setting to the new conditions that a continued rationalization of industry may bring. It seemed as though those concerned, had they obtained understanding for their views here, would not show themselves to be completely negative when it applied to an increase of wages but, in this case, quite naturally, no definite decision was given.[17]

The emphasis on the employers' attempts to retain or increase the freedom of management with regard to how production should be arranged is quite striking. In addition, the association between changes in the general provisions and the pay question appears clearly, as could already be observed in the negotiations 1930/31. The price for an increase in pay would be a concession in the question of the construction of the general provisions and vice versa.

During the following negotiations from 20 November it was clear that this time the big stumbling-block was the construction of the piece wage paragraph. The workers concentrated on preventing unreasonable sackings, on guaranteeing the average earnings whilst awaiting a new piece wage agreement and on safeguarding the provisions about production investigations. The arguments of the employers first aimed at retaining their freedom of action. Job studies were discussed in detail. The main impression gained by the workers' negotiating team from the discussions was that their opponents held fast to their positions, with possibly certain concessions on piece wage bonuses and control of production.[18]

At the beginning of December a new round of negotiations was held and this was concluded by the conciliator putting out a feeler that both parties rejected. Ekblom then declared a breakdown in the negotiations.[19]

The significance of this feeler in so far as it is of interest here was as follows.[20] An increase of 4% was proposed for hourly pay. A piece wage bonus, the size of which was not stated but which obviously was presumed to lie below 15%, was proposed and instead older kinds of bonus piece wages would be deleted. The provision on production investigation was made briefer. No longer included was the provision that the parties jointly should select the worker whose production should be studied nor the provision that investigations should be based on the capacity of an average worker. The workers were no longer guaranteed the right to follow the investigation itself, only to check its result.

In a special minute the idea was developed that different jobs should be graded after the kind of work and weighted between the maximum and minimum limits of a 35–15% excess. There thus returned once again, but now in a more general context, the idea of wage ladders which was acceptable to both parties.

Among the transitional provisions were found again the designs for a piece wage revision. Low piece wages were to be increased up to the 15% level. Piece wages obviously lying too high should, in compensation, be decreased so that they provided an excess on a par with equivalent work in the plant. Even here may be sensed the basic outlines of a wage ladder. As this provision was to be fiercely attacked by the workers' negotiating team there is cause to show how the line of reasoning was developed:

As obviously too high is regarded a piece wage for which the average earnings considerably exceed the average piece wage level at the plant unless this deviation is caused by the type or importance of work. The piece wage earnings-level at the plant is calculated for men and women individually at the highest schedule wage so long as the piece wage does not concern underaged workers.[21]

Here it was a case of a general revision of piece wages not only intended to prune the top from piece wages but also implied a carefully worked out comparison between piece wages at a plant. It is difficult to think that such a comparison was possible without the assistance offered by some kind of job study technique.[22] It was such a thoroughgoing regulation which it was considered prevented building up piece wages in the future that the workers opposed.

A few days later when the negotiation position had been discussed in the general council of the Textile Workers Union, the way in which the negotiating team had proceeded when they rejected the negotiation offer was approved by a good margin.[23] The piece wage paragraph especially had made the negotiations difficult, particularly the clause in the proposal made by the conciliator that dealt with the reduction of piece wages that were too high. The workers were not prepared to go along with a general revision of piece wages similar to that carried out in 1931.[24] But even the bright side of the picture was noted, as that the conciliator proposed in fact both an increase in hourly pay as well as compensation for piece wages.[25] So far as could be understood the employers were prepared to approve the proposal so it was obvious that with regard to the piece wage compensation the ice was at last broken.[26]

Piece wage questions thus played a large part in the negotiations. In their introductory report before the union's general council Wennström had stated:

I can say at once that the piece wage paragraph has taken up the most time. During the discussions the employers have worked on the idea that job studies shall form the basis of all piece wage calculation, but we have then maintained that we should be unable to assess whether the employers' information was correct or not.[27]

In his comments on the feeler put out by the conciliator, Ernst Rosenlund was also interested in job study questions. The minutes state:

With reference to the demand from employer quarters on the right to use job studies, Rosenlund was convinced that the employers would not depart from this demand, but this was a dangerous path for the workers. He mentioned that the employers had invited the negotiating team to a lecture on the job study problem, but they had of course declined this.[28]

The representative from Norrköping, agreed with what Rosenlund had said about job studies.[29] Karin Nilsson, also from Norrköping, had quite a few things to say on the question:

Further, Karin Nilsson stressed that the employer at her workplace (Yllefabriks AB-YFA, author's note) had to give up time measurement because the workers had production statistics with which to meet the employers' arguments. She considered it was very important to safeguard the provisions about production investigations, which the employers use the whole of their energies to get rid of.[30]

It was expected that hourly pay would be increased and they were prepared to accept a revision of piece wages, but job studies must be rejected resolutely. The workers did the work specially nervously in time measurement and could not achieve any result. Many members had left the trade union because the employers had been permitted to go on with time measurement. Karin Nilsson was pleased that the negotiating team had declined the lecture about the job study problem. It was theories that could actually never be demonstrated as correct in practice. Considered it to be of vital importance that the negotiating team maintained the demand to get rid of time measurement, otherwise they would come off badly.[31]

3. The Conciliation Commission's first proposal and the problem of objective basic facts

After the negotiations had broken down on 8 December a conciliation commission was appointed consisting of Anders Wijkman and Hilding Kjellman, both county governors, and Olle Ekblom as secretary. The commission summoned to discussions on 29 December, when the parties only decided to prolong the current agreement until 16 January and to resume the discussions on 7 January 1937.[32]

During January 1937 there were two rounds of negotiations before the conciliation commission, 7–11 and 22–24 January. The breakdown on 11 January, of which more later, resulted in notice of a partial strike from 25 January. Under this threat negotiations were resumed and carried to an agreement on the 24th of the same month.[33]

After the earlier negotiations in January the conciliation commission presented a draft agreement on 10 January to both parties.[34] This differed on important points from the feeler put out by Ekblom a month earlier. Rises of approximately 5% were proposed against the earlier 4%. Compensation for piece wages was now given as 5 and 3 öre for men and women in that order. With regard to the piece wage paragraph a number of changes had been made, which introduced a differentiation of categories into provisional and fixed piece wage schedules. New here were the provisional schedules which had a shorter period of validity than the agreement and did not need to be approved by the main organizations but which were protected similarly by the necessity to give notice in a formal way as the permanent schedules. This was a concession to the employers. The introduction of provisional schedules, to which a tendency was already found in earlier proposals, softened up the rigidity of the piece wage system, which the employers labelled as a danger.

In this proposal the provisions on production investigation were again provided with the same construction as in the current agreement. The shorter version against which the workers had reacted was thus abandoned. On the other hand it was proposed that a minute should now be added to the provision and this was a clear expression of the Textile Employers Association's demand. It read briefly as follows:

Minute. This provision implies no limitation of the employer's right to arrange so-called job studies.

The proposal for transitory provisions in the feeler of 8 December, which would have given the employers the right to decrease piece wages that were obviously too high, had vanished from the proposal of 10 January and been replaced by a provision to the effect that the parties were entitled within 20 days from the time the agreement was concluded to request negotiations and revision of the fixed piece wages. At such a revision, it was laid down, an adjustment should be carried out in the factory: "If the different piece wages within a certain group of piece wages provide an importantly different excess, at the revision an internal regulation of the piece wages within the group should be carried out."
The meaning of the term "group" is elucidated in a minute, where the idea of a ladder-construction has been developed a little further:

Minute
The parties are in agreement that in application of paragraph 7 clause 4 the piece wages at every company shall be graded with reference to the kind of work and its importance. Here the parties shall proceed in such a manner that in the first group they shall place the work that it is considered should produce the highest piece wage surplus. Other work is graded following this into groups with, for each group, a lower excess calculated.

The employers approved the vital points in the conciliation commission's proposal with regard to the provisional piece wage schedules and the texts on production investigations and job studies. But the piece wage compensation was considered too high and for the schedules set out for districts they had a number of detailed objections. The employers' reaction was thus positive and should have permitted a continuation of the negotiations. As however, the negotiating team of the Textile Employers Association heard that their opponents had a much more negative attitude, an adjournment was requested to the following day. The employers then informed the conciliation commission that their proposal was rejected.[35]

That the workers had also been prepared to continue the negotiations on the basis of the proposal put forward by the conciliation commission appears from the speech made by Wennström to the union's general council, which assembled on 12 January. Parts of the proposal could also be regarded by the workers as successes, for example the protection with regard to notice in the provisions about the new provisional piece wages. On this point the workers had succeeded in beating off their opponents' attack. But of course this success did not mean that the conciliation commission's proposal was completely acceptable. The delegates of the Textile Workers Union had without effect maintained that representatives of the local branch of the union should have the right to be present at discussions on piece wages that were regulated in a piece wage schedule. In disputes about such piece wage questions the payment for work was to be made by hourly pay plus 10% until an agreement had been reached. However, on this point according to Wennström, the employers had said they were prepared to accept battle. The workers considered the minute about job studies was unnecessary. Nor had it been considered that the minute about the grading of piece wages on a falling scale should be included in the agreement.

On the points mentioned the demands of the textile workers had not made any headway with the conciliation commission but in spite of this the negotiating team had seen the positive sides of the proposal. What had unexpectedly and radically changed the position had been the Textile Employers Association's suddenly advanced a demand to move two districts, Malmö and Matfors, to a lower wage schedule. This had been completely unacceptable and it had become necessary to break off the negotiations.[36]

Following the breakdown the secretary of the union, J. A. Holm, prepared a communiqué with the assistance of the Trade Union Confederation's press secretary, which explained the workers' point of view and laid the burden for the breakdown in negotiations entirely on their unreasonable opponents. The communiqué contained information about trade conditions in the textile industry and a line of reasoning that worked up to a statement that the actual wage demands from the workers would only increase the wage costs of the industry by 2.74 million Skr, that is to say 0.7% of the industry's production value.[37]

These figures were to play an unintentional but important part as the negotiations developed. The day after the negotiations had broken down the general council of the Textile Workers Union reassembled for discussions. Following Wennström's introductory orientation about the position and the proposal for the agreement rejected by the negotaiting team the chairman for the day, August Lindberg of the Trade Union Confederation, did not immediately declare the meeting open for discussion. He first wished to report that the Employers Confederation's chairman, Gustaf Söderlund, had contacted him in order to obtain clarification about whether the answer from the workers' side to the conciliation commission was ultimative or not. At that time Lindberg had emphasized that the schedule changes for Malmö and Matfors were completely unacceptable. Söderlund had then pointed out that it was completely unfeasible for the textile industry to carry the extra cost, 7 million Skr, that the workers' demand would cost. With astonishment Lindberg had pointed out that the union's secretary had calculated the cost to the considerably lower sum of 2.74 millions. Faced by this great difference Söderlund had declared that he would gladly see a joint comparison of the parties' different figures arranged.

Wennström interrupted the report at this point and expressed his astonishment over the turn the matter had taken. He thought however that the workers would have nothing to lose by such a joint comparison. Lindberg who was obviously strongly positive to the idea

> pointed out that it was not a question of a discussion or negotiations but only purely a calculation problem in order to discover which of the parties' calculations were correct. Or if perhaps both of the parties' calculations were wrong.[38]

Lindberg's report on the comparison of the parties' basic statistics undoubtedly influenced the following discussion about the union's future actions. The joint sitting to look at the figures, which took place on 13 and 14 January, meant a continued informal contact between the parties although the negotiations before the commission had been broken off. Some judges of the situation considered that simply taking the initiative for discussions had shown Söderlund's desire to bring about an agreement and in the continued contact a comparatively mellow, friendly attitude to the negotiations could be observed in Paues.[39] The development of the situation related here provides instant illumination of a number of things in the relations between the parties; things of interest in the development in focus here. First it is striking how the initiative, the seeking of contact, emanates from the spokesmen of the main organizations, from August Lindberg and Gustav Söderlund. Here it is the parties who have been meeting for some months in the revived National Labour Market Committee who react in a threatening conflict between contracting parties lower down in the organizational hierarchy.

The idea from which the representatives of the central organizations were proceeding was that there must be an "objective truth" behind the different information the parties used in their actions and agitation. If the truth could be confirmed, that is to say if the auditors checking up on the figures, employees of the Textile Employers Association and the Textile Workers Union, could agree about the cost calculations, the proportions of the clash in interests would be reduced and perhaps dissolve in the light of the objectively described actual situation.

The pattern of the rationalization movement can be matched with the actions of both Söderlund and Lindberg: what was objectively verifiable should be shown and, assisted by experts, the differences in interests could be reduced to arguments about allocation and suitability.

The concrete question, the possibilities of a statistical base for negotiations accepted by both parties, was to receive a contractual solution ten years later. Then an agreement was reached that gave the Textile Workers Union the right to copies of the primary material on which the other party based its pay statistics. In this way the parties had actually reached a joint, if not objective then intersubjective, acceptance of fundamental facts in the negotiations.[40]

However, this is getting ahead of the situation. The general council decided unanimously on 12 January to give notice of a strike at some factories, which officials of the union and the chairman of the Trade Union Confederation were to select. The notice was not to upset the atmosphere during the comparison of the figures and should therefore be made public first on the night between 13 and 14 January. The call to strike should take effect from 25 January. The long strike notice was decided at the wish of the Minister of Social Affairs, a social democrat, who, according to what was said, sympathized with the workers' demands and wanted the chance of urging the commission to issue a call for new negotiations under the threat of conflict.[41]

4. Agreement

In a communication to the Textile Employers Association the Textile Workers Union gave notice of a strike at 5 companies. In the days that followed the employers also received similar communications from the other trade unions involved.[42]

The idea of course was that using the threat of conflict the union would force the conciliation commission to recall the parties for negotiations, and this occurred. New negotiations took place. On 24 January after two days of discussion both parties signed an agreement in accordance with a draft put forward by the commisssion.[43]

The agreement was based on the prolongation of the provisions in the earlier agreement but with comparatively large alterations in paragraph 7 (the piece wage paragraph), which was given a completely new wording. In a comparison between the final agreement and the proposal made by the conciliation commission early in January it is clear that it was the employers who had to make the most important concessions in the last round. Certainly the possibility of provisional piece wage schedules was obtained. Approximately another öre had been added to the schedule wages so that the increase in relation to the earlier wages was 6%, and piece wage compensation had been increased somewhat (earlier 5 and 3 öre, now 7, 6 or 5 öre). The right of the workers to consult the local trade union on piece wage agreements was admitted in the agreement. With regard to the remuneration to be paid whilst waiting for a piece wage the workers made some advance but did not achieve their goal. The new agreement laid down that if the management, for example in connection with rationalization measures, wished to delay the offer for a new piece wage, for work normally carried out at a piece wage, the workers should have the right to hourly pay +10%. In disputes about piece wages, on the other hand, the work should be carried out until further notice at the offered piece wage or at hourly pay.

The minutes proposed in the earlier request about job studies and "ladder construction" had been deleted in the text of the final agreement. But there still remained the provision that if a revision of piece wages was requested by either party, attention should also be given to the adjustment of the piece wage between the different jobs.

5. Collective negotiations 1938

Following the clash at the beginning of 1937 the negotiations on the national agreement for the textile trade became calmer for a long period. The agreement concluded in February 1937 was for two years so it was not until the summer of 1938 that the question of the agreement became actual again for the Textile Employers Association and the Textile Workers Union.

The general council of the union decided as early as May 1938 to try to arrange preliminary negotiations and during the latter part of September to convene a policy conference to discuss the agreement in order to decide about the possible result.[44] The union's leadership was also keen to coordinate its own collective negotiations with the other members of the Trade Union Confederation. At the Trade Union Confederation's joint policy conference at the middle of August the confederation's leadership had urged some degree of care and moderation in the demands put forward by the trade union movement.[45]

The day after the conference Gösta Wennström contacted the Textile Employers Association and agreed to start preliminary negotiations at the beginning of September.[46]

At these the workers put forward proposals for changes in the agreement, which in line with the recommendations of the Trade Union Confederation were marked by a certain degree of restraint. In any case it was not possible with the frame work of a prolongation of the agreement to gain any particularly great new advantages. The textile workers demanded a wage increase of the same size as that once given by the agreement then in force, that is 6% on hourly wages and an additional one öre on piece wage compensation. The piece wage excess for existing piece wages was to be set to at least 20% when, however, the limit of 15% would be accepted in the new settings (where the building up possibilities were still not exhausted). Finally was demanded a change in clause 1 c of the piece wage paragraph: in the delayed setting of piece wages the hourly pay bonus should be increased from 10 to 15%.[47]

There were two rounds of negotiations during the first half of September.[48] In the situation as it then was it is clear that in their internal discussions the employers thought they were on the defensive. No coherent proposal was presented by the management of the Textile Employers Association, which undoubtedly intended only to try to gain a two-year prolongation with the least possible concessions to the demands of the workers. It is clear that among the concessions the association was prepared to make from the beginning there was a certain increase in pay, the size of which was not stated, and also adjustments on a number of other points, not including those parts of paragraph 7 of interest in this context.[49]

During the second round of negotiations at the middle of the month both negotiating teams were able to agree on a draft for the agreement. This implied prolongation of the current agreement but with certain alterations. The hourly wage schedules were to be adjusted upwards so that the total result became an increase of approximately 6.5 %, allocated with the greater part on the women's schedules and the lesser part on the men's. The bonus in a delayed setting of piece wages was increased to 15 % of the hourly pay which of course was also the lowest target value in the setting of new piece wages. It was confirmed in the minutes that if a worker was given more machines to look after the piece wage should automatically be terminated. All piece wages were to be revised so that they gave at least 15 % above the new hourly wages, calculated on earnings during the first half of 1938 or on earnings during the period the piece wage had been in operation.[50]

When the organizations had each decided to approve the settlement it was possible to conclude the agreement on prolongation with smaller alterations before the old agreement ran out.[51] The agreed revision of piece wages, however, was still not completed at the end of March 1939.[52]

6. Summary

The second half of the 1930s also started with a greater clash between the two sides of the textile industry. But the differences of opinion this time did not escalate further than the issue of a strike warning that was never put into force. In a comparison between the collective bargaining of 1930/31 and 1936/37, it is also worth noting that the opposition in the Textile Workers Union of the earlier period had no equivalent in the latter, if the central material is anything to go by.

During the collective negotiations of 1936 it was the workers who were advancing. Mainly the workers on hourly pay were looked after by the introduction of piece wage compensation; but the increases in the schedule wages, from which all categories gained, were important successes. Certainly the workers accepted a revision of piece wages in accordance with the new customary pattern so that the highest piece wages were reduced, but the workers took a stand against the idea of fixed steps between the piece wages for different work, which had been expressed in numerous ways during the negotiations carried on. The wage ladders that became the further development of this idea were to belong to the 1940s.

In the case of the procedures for piece wage agreements, there occurred with the 1937 agreement an ingenious shift in the balance of influence that gave both parties some degree of satisfaction. The employers could observe that the total locking of piece wages, threatened by the permanent schedules had been broken by the possibility of provisional schedules. This was a way for the employers to safeguard the local freedom of action of management. But the local trade union organization obtained an increased chance of influence as the agreement guaranteed the individual worker's right to consult his union. The tendency of the fixed schedules towards rigid centralized procedures was broken and replaced by the advantage of a wider freedom of action and an increased influence for both parties at the local level.

The last peacetime collective negotiations during the 1930s certainly brought about a further increase in the earnings-level but otherwise the construction of the agreement remained unchanged. One important addition was made in the shape of a minute to the agreement: transfer to more machine units, perhaps the most frequently used expression of rationalization, should normally lead to negotiations about a new piece wage.

CHAPTER 12

Collective negotiations in the 1940s

1. The careful years

The national collective negotiations carried on at the beginning of the 1940s between the large parties of the textile industry were obviously affected by the stresses of the war. A pervasive characteristic was the association with the cost-of-living index agreement concluded between the Employers Confederation and the Trade Union Confederation. This agreement was so constructed that space for wage increases was made dependent on changes in the cost of living. If the cost-of-living index exceeded certain stated intervals an index bonus was granted on hourly wages equivalent to defined percentages calculated on the wages for 1938. The index agreement was naturally respected but could be considered unfavourable to the workers in low income industries: for the lower wages had been in 1938 the lower was the automatic compensation awarded by index agreements for the rise in prices. Needless to say, this state of affairs constituted one of the textile workers' arguments in negotiations. The authority of the central index agreement grew from 1942 when a price and pay stop with a general effect was introduced. This started to lose its effect in the autumn of 1944 as the beginning of the end of the war could be seen and because both agricultural workers and the lower levels of civil servants had received pay increases of 11–12% in the earlier pay negotiations.

Negotiations on the textile agreement were ususally carried on as preliminary negotiations or during short-term provisional extensions of the current agreement.

In the negotiations prior to the 1940 agreement the workers advanced claims for increases in the piece wage compensation (expressed in öre), which should also apply to workers who had reached 20 years of age. The current agreement laid down 21 years as a minimum limit. Piece wage conditions for workers below this age were also brought up for discussion.[1]

The agreement concluded at the beginning of 1940 rejected the demand for changes in the basis of piece wage compensation. But the parties agreed on the revision of piece wages that gave an increase in the price per piece. The minimum level of 15% which had been applicable for 10 years was retained. However, no reduction to the 35% level was presumed. Instead, piece wages that should give a lower excess than 35% were increased by 4 öre, those that gave an excess between 35 and 43% were to be increased by 3 öre and, finally, those that gave an excess between 43 and 50% were to be increased by 2 öre. As the upper target the 35% limit for piece wages had thus played its part in the national agreement.[2]

The national agreement was also extended for 1941 without any special alterations or demands for alterations by either party.[3] It is perhaps worth noting here, as showing the assessment of important conditions by one of the parties, the wish expressed by Gösta Wennström when he reported that the workers approved the proposal submitted for settlement:

> It should further be desirable during the present time of crisis with continually increased costs of living that the employers do not unnecessarily by the application of paragraph 7 clause 1 b of the national agreement terminate the present piece wages because of relatively unimportant changes, with the purpose of decreasing earnings.[4]

Apparently Wennström still considered, as in the mid-30s, that the rationalization clause was misused by the employers in certain quarters so that "rationalization" only became an excuse for lowering the current piece wages (cf. p. 242).

But the compliment was to be returned. At the beginning of the following collective negotiations, which applied to the agreement of 1942, Otto Thurdin, who was in practice manager of the Employers' association, delivered a comparative charge against the workers in similar terms:

> The employers also wanted to refer to the situation that unfortunately it still occurs that the workers with the intention of misleading in the setting of piece wages intentionally restrain their working speed. The leaders of the Textile Workers Union had lately at various negotiations on such questions dissociated themselves from these tactics but nevertheless the employers would like to express the hope that the Textile Workers Union would still further stress to its members the wrongness and even, from their own point of view in the long-term, the uselessness of it.[5]

At these negotiations neither party requested changes in the piece wage paragraph. The Textile Workers Union again put forward the demand for a lower age limit for piece wage compensation, which also made its way into the agreement concluded at the end of January.[6]

When the parties met again in the autumn of the same year both sides had accumulated demands for changes that also concerned the piece wage paragraph. The workers put a great deal of weight on being able to increase the minimum limit for the piece wage excess. At the beginning of the negotiations a minimum target of 25% was requested. Additionally requested was that the payment in the delayed setting of piece wages should also be raised to 25%.[7] As day follows night the employers replied with a reference to the newly proclaimed general pay stop. That the whole corps of textile workers shold be an exception. as Wennström had maintained, was an argument the employers found it impossible to accept.

The Textile Employers Association also wished changes to be made in the piece wage paragraph. In the "rationalization provision" they wanted to exchange the expression "with regard to more machine units", which was stated to be misleading, with "with regard to a high degree of effort or work". Here it was a question of a countermove against the minute in the agreement of 1939. By using the second wording the tending of an increased number of machines would not be suffiecient to cause new piece wage negotiations, but it still had to be shown that the change carried with it in fact an increased burden for the worker, which was of course a great deal more difficult. At the same time such an assessment would make more necessary, or give advantage to, the side that controlled the job studies.

In the provision concerning production investigation the employers wanted to include still another condition. It should not be sufficient to use as a basis for piece wage setting the production that *on average* was carried out by an *average* worker. In addition, said the employers, the condition should be fulfilled that *normal* conditions should apply and that the investigation should be based on the production results of a number of workers.[8] The meaning of this proposed change was that the basis of calculation was withdrawn from the simple observation of a worker's actual production to somewhat more advanced statistical calculations of normal values at a more theoretical level. Later during the negotiations the employers made additions to their demands that also included a request "that the workers' side will assist in measures for the equalization of fixed piece wage schedules that have an uneven effect".

The negotiations, in the final phase conducted before Olle Ekblom as neutral chairman, again resulted in an extension of the general provisions of the agreement, with the ususal increase in the low piece wages to the 15% limit and a smaller addition to the hourly wages in the lowest schedules.[9]

The workers returned to the negotiations on the 1944 agreement with a list of demands maintaining the same actual claims with regard to piece wages, but at a somewhat lower level of ambition. The target for piece wage excess, payment in the delayed setting of piece wages, and piece wage revision was stated this time as 20%. An addition of 6 öre on hourly wages was also demanded.

This time the conflict of opinion was tougher than it had been during the last few years. Hilding Kjellman took over as the neutral chairman. In many rounds of negotiations it could only be noted that the parties were far from agreement. The greatest differences were caused by questions other than that of the piece wage problem. At the beginning of January Kjellman decided it was time to break off negotiations, whereupon the workers gave notice to terminate the agreement. A conciliation commission was set up where Kjellman was assisted by Olle Ekblom and Per Westling.

After only a few days of negotiation the commission succeeded in getting results. An agreement was concluded that meant considerable increases in wages for hourly workers, including an increase of 2 öre for piece wage compensation. Also included in the agreement was the customary provision on piece wage increases to the minimum level.[10]

2. The workers' harvest

In the period leading up to the negotiations during the autumn of 1944 the crippling effects of the war had continued to slacken. When Gösta Wennström presented the workers' draft agreement his speech was essentially marked by a much more acrid offensive spirit than in the earlier years of the war. The textile workers' chairman could show the increased pay of agricultural workers and the lower grades of civil servants and was able to maintain that the textile industry had gone through the war with good profits. These assessments led to the demand for piece wage excess and payment in the delay of piece wage setting of at least 25 %, regulation of the cost-of-living index to proceed not from the wage schedules of 1939 but from those of 1943, an additional 10–12 öre addition to hourly wages and an increase in piece wage compensation of 5 öre. On the other side of the fence Otto Thurdin pointed out that the previous year's agreement had become costly and maintained that the textile workers had already achieved the goal set up by August Lindberg at the negotiations between the Employers Confederation and the Trade Union Confederation in the autumn of 1943: to get back to the real wages of 1939.[11]

Nevertheless it is quite obvious that the workers were supported by the economic development: nobody denied that the textile industry had shown good financial results. The favourable position of the workers in the negotiations, conducted this time under Hilding Kjellman, did not however find expression in changes in the piece wage paragraph. Instead it was the hourly paid workers who made gains. All schedules were increased by approximately half the amount the workers had requested and the period of qualification for piece wage compensation was decreased from one year to three months. The parties also agreed about the customary rise in low piece wages to a 15 % excess. Otherwise no demands on further increases in piece wage earnings were permitted.[12]

The collective negotiations preceding the agreement of 1946 followed the pattern of the recent ones. At a rather early stage the direction of the negotiations was taken over by Hilding Kjellman. The workers took the offensive and proceeded with largely the same demands about piece wages as in the previous year.[13] After a number of rounds of discussion a new agreement was concluded at the beginning of February.

The new agreement meant an emphatic increase in hourly pay of 10 öre and an increase in piece wage compensation in the highest schedule. The piece wage revision followed the design of the agreement at the beginning of 1940.[14] All piece wages were raised to a minimum 15% in the usual way; at the interval 15–25% they were increased by 7 öre and for the interval 35–50% the increase was 6 öre. The piece wage excess of 50% was now regarded as a ceiling: there were to be no increases in addition to these.

The settlement also reflected for the first time the reality that began to rear its head out in the companies although the idea had been rejected in connection with the agreement of 1937. In a minute it was stated that in those cases there "wage ladders" appeared all targets should be increased just as much as in the general hourly pay increases.

The "wage ladders" were local agreements about fixed piece wage targets for work that demanded different training or had different degrees of difficulty. In a further minute the parties undertook to work for the adjustment of internally uneven piece wage setting within each individual company. In this way the appearance of these locally fixed earnings-levels were adopted into the textile industry agreement. This had the effect, as earlier authors have observed, that both sides also accepted the careful work-evaluation aiming at objectivity on which the "steps" were constructed. In their turn these could hardly be considered to apply in practice without the help of job studies. The entry of the wage ladders into the textile agreement thus also means final acceptance of the part played by job studies in the textile industry.[15]

The conditions before the agreement of 1947 had altered on certain points.[16] As in the previous year the textile industry still enjoyed a boom and undoubtedly this carried with it the hope of a continued harvest for the textile workers. Concurrently, through an agreement between the Employers Confederation and the Trade Union Confederation, the end had come for the system of the war years with an automatic link between the cost-of living index and a crisis bonus. The collective negotiations betwen the parties of the textile industry were thus able to proceed with greater independence than for quite a long time.

As in the previous year the negotiations were carried on under the chairmanship of Hilding Kjellman. There were a number of rounds between December 1945 and February 1946. The workers made claims for considerable improvements: increases in hourly pay of over 20 öre, and increases in piece wages of the same order of size or more so that the lowest piece wage excess was 25%, etc.

There was no doubt the employers were on the defensive. In the final agreement, however, it was clear that the Textile Employers Association in exchange for an increase in schedule pay of 13 öre (for adult male workers, to which there was an adjustable bonus of a further 5 öre) had managed to obtain the removal of the automatic relation between hourly pay and piece wages. This was achieved in two ways. One was that piece wage compensation disappeared as a special conception in the agreement by being included in the schedule wage increases. At the same time the workers had to undertake not to raise the idea again at the next collective negotiations. The second was that the piece wage target for average workers with normally difficult work was no longer expressed as a percentage (that is to say, at least 15 %) of the hourly wage. Instead a target in öre was given for every wage schedule. The actual table was, however, still based on values that corresponded to approximately 15 % above the hourly wages.

Through this separation of hourly wages and piece wages the employers were undoubtedly better equipped to deal with the continued offensive from the workers; an offensive whose aim often showed itself to be directed at the lower income groups, the situation with regard to the earnings of workers on hourly pay. Such concentration on low wages in the name of a unified wage policy should not automatically also bring about changes on the piece wage side. At the same time the piece wage targets in öre undoubtedly constituted a high degree of standardization in piece wage setting over the whole of the textile industry. The construction may therefore be seen also in the light of the employers' attempts to deal with the wage drift which had been increasingly more troublesome because of the scarcity of labour.

3. Summary

The index agreement between the Employers Confederation and the Trade Union Confederation quite obviously came in both its construction and content to influence the negotiations and agreements of the 1940s between the Textile Employers Association and the Textile Workers Union. But there was still some room for further demands from the workers. A continuous and especially at the beginning of the period, dominating demand was for increased earnings for the groups of workers on hourly pay. This was to be achieved by increased values for piece wage compensation over and above the schedule wage increase received by all. From 1943 when the offensive spirit with its optimism for the future returned, an increase in the piece wage targets was demanded regularly and fruitlessly. Concentration on increases in hourly pay was of course correct from the workers' point of view. Those paid by the hour formed a low pay group in the textile collective and the concentration could therefore provided with an ideological foundation that added strength to the demand. Furthermore, the increases in schedule wages also brought about increases in piece wage earnings as the targets for the latter were expressed as a percentage of the former. It was also this circumstance that brought about the construction with special piece wage compensation for workers on hourly pay, an addition that did not affect the basis of piece wage setting.

In the agreement of 1947 piece wages were separated from schedule wages by expressing the targets for the piece wage excess in öre. This was a measure with initially a centralized implication. Piece wages were normalized in the national agreement instead of being made dependent on the local wage agreements. One advantage of the new procedure for the employers should have been a greater control over the wage drift of piece wages, a problem noted in the internal discussions of the Textile Employers Association. For the association centralization also meant a better chance of control and influence, but in this case mainly in relation to the opposing side. The freedom of piece wages was reduced.

Dissolution of the relationship between piece wages and hourly wages made it possible to simplify the construction of the agreement. Piece wage compensation as an independent conception disappeared and its value was included in the hourly pay schedules.

During the 1940s the idea was again revived of a classification of piece wages into grades, which has already been brought up by the technical committee of the Textile Employers Association in 1936. Often the revisions of piece wages took the form of general increases of different sizes depending on how high the piece wage excess was. The de facto division into grades of the piece wages naturally followed from this at the various plants. The employers tried to obtain a normalization on this point so that the uneven piece wage earnings should be smoothed out. The workers opposed this up to the agreement of 1945 when a minute to this end was made. With the introduction of this minute the Textile Workers Union had in fact also accepted the idea that an objective and fair wage-ladder could be constructed and thus admitted the means for the practical application of such a construction, namely job studies. During the 1930s the employers protected the right to use job studies in the setting of piece wages against the determined attacks of the workers; at the middle of the 1940s provisions were introduced into the agreement that clearly presumed job studies as a basis for piece wages.

CHAPTER 13

Independent arguments of the parties

1. Introduction

In the earlier analysis the attitudes to job studies expressed by the parties of the textile industry in collective negotiations and central disputes have been analysed. In such situations the reactions of the organizations were naturally dependent on consideration being given to other matters and to the frames of reference that existed, not least because of the comparative strengths of the parties. Nevertheless, in spite of this limitation it has been a basic factor in the present investigation that the actions of the various associations involved in such locked situations reflected not only the limitations but also the targets for their activities.

This clearly supplies a reason for filling in the picture by considering other kinds of expressions of opinion, more or less independently formulated statements on policy. As will be seen, such statements about time and motion studies were comparatively rare from the parties of the textile industry but some examples exist and will be dealt with in this chapter. This makes possible a comparison between how the organizations formulated their basic attitudes and how they acted in negotiation situations when freedom of action was limited by many considerations.

2. Line taken by the Textile Workers Union

As was made clear earlier rationalization within the textile industry accelerated during the latter part of the 1920s and into the 1930s. It has also been seen how the workers in the negotiations of 1930 tried to limit the chances of the employers using job studies as a basis for the setting of piece wages. These two circumstances show that job studies became an increasing problem, in some cases a threat, for the workers in the textile industry as for certain other groups of industrial workers.

Some two years into the 1930s the Textile Workers Union's leadership worked out a line in the job study question that could be introduced into the political work of the union. In 1934 and 1935 Ernst Rosenlund, a representative of the union, contributed articles to Textilarbetaren (The Textile Worker) in order to show clearly the attitude to job studies sanctioned by the leadership.

The first of his articles had the general title of "Rationalization" and may be regarded as the adoption of a stand with regard to more overall rationalization activities. At the same time it is clear that the article was brought about by certain circumstances that Rosenlund refers to in more or less clear terms. One of his points of departure was from a motion in the Riksdag, the Swedish parliament, which was submitted in the spring of 1933 by four social democrats. Rosenlund wrote:

> In our opinion society must in future more than up to now give its attention to problems in this category. It is with pleasure one notes that in this year's Riksdag a motion has been introduced about the investigation of rationalization and its aftereffects and we very much hope that this legislative assembly will give the question the attention due to it.[1]

This text referred to a motion in which was demanded that society should carry out an investigation of the negative social effects of rationalization especially with regard to unemployment in order to consider if the public should be given greater influence on the rationalization of industry.[2] Rosenlund's reference to this motion helps to date the article. It must have been written during the spring of 1933, before the 13 May, when the second chamber decided not to agree with the mover of the resolution. Thus the article should have been written a long time before publication. This had the result that Rosenlund, when the Rydboholm dispute about job studies made it natural, was able to bring his text up to date by simply adding a column that implied an attitude to the Bedaux system.

The idea that the article was written in stages is also suggested by its disposition. Initially rationalization was dealt with in a historic perspective. Rosenlund observed that the textile industry had always been well to the fore in modernization. He referred to the English Luddites, who at the beginning of the 19th century smashed the mechanical looms they regarded as a threat to their own chances of earnings. Rosenlund took a stand against such reactions. Following this he presented his view of what rationalization had meant:

> The amazingly rapid rationalization we have become acquainted with must quite naturally lead to irritation and friction. The employers utilized the technical improvements solely for their special purposes. As soon as labour-saving machines are introduced or more machines are given to every worker the employers have taken the opportunity to force down piece wage pay. This method of procedure must obviously carry with it bitterness and resistance from the workers' side.[3]

Rosenlund admitted that the employers had a contractual right to terminate piece wages when production methods were changed but made it quite clear that the "rationalization clause" was misused and emphasized that the new piece wages must result in a sharing of the profits of rationalization, thus compensating the workers for an increased working pace and additional work. In the work of rationalization the employers did not consider the social aspect, said Rosenlund, and once again reminded his readers about the Luddites' rampages. Who could guarantee that the workers of the 1930s would not react as their forerunners had done if conditions were not improved? In order to gain improvements public means must be utilized. Among the concrete measures suggested by Rosenlund were shorter working hours, prohibition of shift working for minors and the establishment of unemployment insurance in the form of a pension.

It would have been quite natural had the article finished with this list of demands. But Rosenlund took a deep breath and went on to discuss the Bedaux system:

> In their zealousness for profit and a love of all kinds of fleecing systems, these employers have not cared a fig for current agreements and the workers' ideas about the suitability of introducing the system in question. In our opinion one cannot obtain more convincing evidence of the employers' egoistic view of rationalization than in this case.

Without in fact describing the Bedaux system in other terms than a system for the control of work and the calculation of wages, Rosenlund used the system for a new broadside against the employers and the way in which they carried on rationalization. Rationalization, he said, had come to be regarded by the workers as a "Pandora's box" out of which flowed all kinds of miseries.

The article concluded with a future vision of a society controlled by the workers' movement where rationalization had become also for the working class a weapon used in the battle for welfare.

The principle view essentially adopted by Rosenlund was that current among the social democrats; a conditional attitude to rationalization if the necessary social consideration was given and social reforms carried out so that the workers should be able to deal with the increased rate of working. This was a relativistic attitude spiced with tones of a more radical kind where the Luddites' shadows emerged from the depths of the past and where the real blessings of rationalization were made dependent on the shaping of the social system.

Immediately in the following number of Textilarbetaren rationalization was dealt with again. This time in a detailed report from the International Congress of the Textile Workers. Among other questions on its agenda the congress had to decide about rationalization. In a commission given the task of producing a statement on this point was included the Swedish representative, Gösta Wennström. The statement made by the international congress reflected a view that was also supported by the Swedish organization.

The statement derived its character from the following opening where the positive about rationalization was contrasted in four points with the negative, also in four points:

The 14th International Textile Workers Congress does not oppose rationalization methods which
1) improve methods:
2) produce perfect production methods:
3) increase the sales of goods: or
4) by means of scientific working method increase production at the same time as they decrease tiredness in the workers and thus increase their health and well-being, so that the work becomes more pleasant for them.

We oppose however most definitely, as before, the introduction of working conditions by means of which
1) the workers are converted into nothing more than machines, which carry out their work automatically without needing to think;
2) people like the revolutions of a wheel are controlled by means of a clock and the human element completely removed;
3) in a full day's work norms are used, which are only applicable to concentrated work for shorter periods of time;
4) physical and spiritual tiredness is increased.

The latter methods, which in our opinion were incorrectly described as rationalization, we shall oppose with all our might and fight them with all the means at our disposal.[4]

Following this preliminary an argument was put forward in which as a starting point it was laid down that all rationalization should proceed from the working ability of the workers. This alone passed judgment on all changes that increased the burden of work, efforts and tiredness. As a problem in all the countries concerned (the international congress included only the workers of the European textile industry) appeared especially the tendency to supply workers with an increasing number of machine units without corresponding relief. Rationalization was also rejected as a means by which it should be possible to shake off the world crisis. (In a similar resolution on the shortening of working hours a further step was taken: technical development was pointed out as a cause of the world depression). At the end of the resolution on rationalization it was stressed that only by governments in the industrialized countries increasing the purchasing power of the consumers could the balance in the economy be regained. This was certainly a remarkable stand to take by a trade union organization, to advise governmental measures so unhesitatingly as the only way out of the crisis. It would have been natural for the trade union struggles to have been shown more clearly in the context. It was also worth noting that the resolution concentrated so wholeheartedly on conditions of industrial safety, working hours and the speed of work, while the question of sharing the profits of rationalization was not touched upon at all. This confirms the impression that the statement was less concerned with drawing up the lines for trade union action than to stress the need for political measures.

During the following year, 1935, Ernst Rosenlund returned to the questions bound up with rationalization and especially to the greater of the problems to which he had devoted the former article, job studies.

In an introductory discussion the author drew attention to the use of job studies and linked this to a desire of the employers to increase the intensity of work independent of the negative consequences that might arise as a result of this activity that was governed by profits. Rosenlund also associated this demand for a more intensive working speed to the shortening of working hours that workers in most of the European industrial countries had achieved after the war. He remarked on Taylor's influence and mentioned the work carried out by the German Refa, Reichsausschuss für Arbeitszeitermittelung, which served as a model for the time and motion study commission of the Swedish Academy of Engineering Sciences.[5]

What benefit was this work of rationalization? Rosenlund moved on to an evaluation and criticism of the development:

> However, we consider that too much time and money have been sacrificed and is being sacrificed, in this matter. Both of these could certainly have been used better for more useful purposes. The employers may think how they like but it can never be proved that industry has had any real advantages from the job study system.[6]

But it was not enough that job studies were a waste of time, they could in addition often be quite injurious. There were many examples of unsuitable job study men in Rosenlund's opinion. It could be demanded that they themselves knew how to do the jobs they studied and had put themselves in the workers' conditions on the whole, so that they could understand the extent of the effects of job studies. But this was not always the case.

To utilize job studies as the basis for the calculation of wages was described by Rosenlund as "absurd". There was also the fact that the employers by means of unilateral decisions tried to introduce these methods of calculation:

> The workers have not even been asked for their opinion about the 'studies' in question. In any case the employers in the textile industry are so conservative that they have not considered it suitable even to discuss the matter with the workers' organizations impartially. We have discovered that the employers in other industries have made *agreements* with the workers about under which conditions job studies shall be carried out and for what purpose the results of these shall be used. This may be so—the employers in the textile industry had better keep their faith in job studies to themselves. Through the national agreement the workers are protected against the effects following a misuse of the job study system when it is a case of setting a price for piece wages.

Apparently here Rosenlund referred to paragraph 7 clause 3c of the agreement, the passage about production investigation pushed through by the workers during the last collective negotiations. It appears quite clearly here that the workers' side regarded, or wanted to regard, production investigations, as a protection against the technically advanced job studies. With production investigation the workers thought that the starting point for the calculation of piece wages was obtained through the actual production statistics, which were easy to understand, which could be studied afterwards and were available to both parties. Such a statistical investigation was possible for the workers to accept as a basis for the setting of piece wages but not the job studies' almost impenetrable synthetic "on the dot-finished check".

It is obvious that Rosenlund in this extrovert attitude to job studies followed the line advocated by the union in the negotiations of 1930–1931.

Even elsewhere in the union than the executive board the job study problem was taken up. The problem was discussed at the meeting of the representative council in the spring of 1936. Karl Nilsson from Halmstad pointed out that job studies were used in most countries and the job study men often behaved "in an irritating way" towards the workers when they tried to push them to the greatest possible performance. He considered that the psychological pressure on the female workers especially became too great. Therefore the unions should provide all possible help to the local branches in job study problems.

Gösta Wennström defended the union: certainly the branches had been provided with the support these might need. In some companies the union members had in fact the right of joint determination with regard to job studies, as at Norrköping. But when it was a case of using the results of these studies as the basis for the setting of piece wages, the point of view of the union had been clear for a long time: job studies could not be utilized in that manner.

Ernst Rosenlund concurred with this argument and recommended that the branches keep an eye on how the "rationalization clause" was observed. They must make clear to the members that they were not obliged to accept termination of piece wages other than on the grounds stated in the piece wage paragraph of the agreement, that is only if real changes had been made in the conditions of the work. Instead of job studies as a basis for piece wages they should request production investigations in accordance with clause 3 c of paragraph 7 of the agreement.

Even at the congress of the Textile Workers Union in September 1936 job studies were brought to the fore, this time through a motion.[7] This was a personal motion and so not approved by the branch.

Behind the motion lay indignation over the way in which job studies were used. Time studies entailed such a pressure on people in production that they became worn out before their time if the working hours were not reduced also. There was also the fact that when piece wages were set on the basis of the artificially raised performances and then calculated to give the minimum excess of 15% laid down in the agreement, this meant in practice that it was not possible in practice to achieve the earnings level the agreement was intended to guarantee. The motion concluded with a request that the union should in the coming collective negotiations demand such wage provisions

> that protect the workers against injustice in the carrying out of job studies and guarantee them the consideration that they as workers have the right to demand.[8]

The executive board's statement, which was approved by congress, may be said to be the union's first official stand with regard to the use of job studies in the textile industry. By way of introduction the board confirmed that job studies had been misused in certain areas covered by the union. In the assessment of the consequences of these studies the unanimity between the leadership and the mover of the resolution was complete. This view of the effects resulting form job studies further caused the board to deny them any evidential value as a basis for the setting of piece wages. Opposition could be expected on this point in the approaching collective negotiations: it could be presumed that the employers wanted to bring in provisions favouring job studies, but the workers' line still stood firm, to submit instead that piece wages should be calculated by production investigations controlled by both parties.

If the union, as now proposed, started negotiations about the contractual regulation of the use of job studies, quite apart from the fact that the purpose was to protect the workers against misuse, this could be interpreted as though the union had abandoned its earlier point of view. This would of course imply an implicit acknowledgement that job studies, presuming that they fulfilled certain demands in their operation, could in fact be used as a basis for the setting of piece wages. In order to avoid this risk the union should keep to its earlier view, at least to lie low for the time being and limit itself as "up to now" to carefully follow the development.

In spite of being urged by Ernst Rosenlund to take the opportunity for a discussion on job studies so that the leadership could get an idea of the feelings about them out in the branches, nobody rose to speak on the motion. Without any debate the board's statement was approved.

Later in the autumn of the same year the chairman of the Textile Workers Union had the opportunity to give his views on job studies in connection with the conference of the Rationalization Committee[10] (cf. p. 153). Not unnaturally the assessments given by the workers' representatives emphasized the negative sides of job studies—their importance and utility had already been stressed by the employers.

The secretary of the union, J. A. Holm, considered that job studies were given a bias to increase working speed and were only used to a lesser degree to really locate faults in the manufacturing process. Possibly, he admitted, there was here a difference between small and large companies as method study should be more developed at the latter. To a certain extent Gösta Wennström spoke against his own colleague when he showed how job studies were often directed at changes in method, but only with the purpose of being able to terminate a piece wages with the excuse that the basic conditions for the price had altered. Because of this the workers did not receive their just share of the fruits of rationalization. The negative aftereffects of method studies were also noted in another way, he said. It paid to make production cheaper by decreasing the workforce. For the workers this meant sackings and the threat of unemployment.

The Textile Workers Union's view of job studies at the middle of the 1930s is well documented both through its actions in the collective negotiations and the statements made in various official and unofficial connections. No comparative formulation of policy was made later by the union during the period covered by this investigation. The characteristic, basic features are clear. The Textile Workers Union wholly rejected job studies as a basis for the setting of piece wages, the union adopted a suspicious attitude to these studies because there was a risk they would lead to unemployment and an increased rate of working, and also because they were misused in order to justify reductions in piece wages.

3. Line taken by the Textile Employers' Association

It is more difficult to discover such general but at the same time sufficiently specific statements about job studies, which it would be possible to call statements of policy by the employers. This was partly due to the nature of the problem. There was not the same reason for the employers' organizations to make public statements on the question. To the degree that job studies and their associated problems dealt with the setting of piece wages and other pay problems, they were part of the organization's regular approved activities and in those matters that dealt with method development they were, according to the Textile Employers Association's way of regarding the matter, a subject for the individual company with which the organization should not interest itself. This was a logical result of the division into negotiable allocation questions and non-negotiable production questions, which for a long time had been fundamental for the regulation of the Swedish labour market.

The relative absence of verbalization obviously did not mean that the representatives of the Textile Employers Association took no stand on job studies. Their behaviour in the collective negotiations of 1930 and 1936 exhibited a firm and resolute defence of the right to the job studies considered unavoidable. The same assessment was put forward by the association's representatives at the conference arranged by the Rationalization Committee mentioned above. Representatives of both the woollen industry, August Schmitz of Furulund, and for the cotton industry, Axel Bergengren of Borås Wäfveri, stressed the great and important role of job studies for the Swedish textile industry. The latter maintained that methodical job studies had been carried on to a great extent in the industry under the direction of specialized personnel trained by engineers. Schmitz raised the objection, however, that there was presumably a difference between larger and smaller plants in this. No doubt he was correct. Bergengren's experience from the purposeful work of rationalization at Borås Wäfveri, where the rationalization pioneer Alex Engblom was appointed as chief engineer in 1920, was hardly representative for most companies.[11] One confirmation that time and motion studies were being awarded an increasingly greater role in production and thus also for the shaping of the Textile Employers Association's agreement policy is the birth of the so-called technical conference of the association (cf. p. 215). With the introduction of such studies it naturally followed that the foundations of the relations between management and workers rested on technical conditions to an accelerating degree. This of course caused difficulties for the managers and lawyers who met together in the leadership of the Textile Employers Association. The technical conference was first assembled in order to draft the questions prior to the collective negotiations of 1936–37 and then became in practice the permanent drafting body of the general council on questions of a technical and managerial character. In the conference were included people on the "chief engineer" level, recruited from the highest technical management of the big companies.[12] Comparison with the Metal Trades Employers Association lies easily to hand.

The adoption of an attitude and evaluation of time and motion studies turns up at the latter part of the period investigated. When the Labour Market Council took up the problem of time and motion studies in 1946 (cf. pp. 258 et seq.) the Employers Confederation requested opinions about a memorandum on a contemplated joint body for collaboration. In its answer to the main organization the Textile Employers Association stated their ideas about the actual situation, and also a number of programmatic formulations and recommendations.[13] These will be reproduced here in some detail.

The author of the memorandum maintained that time and motion studies had played a great role in the attempts at rationalization that had characterized the Swedish textile industry ever since the 1920s. In fact during that period of the 1930s when productivity had increased extremely vigorously, time and motion studies were utilized intensively. In another context it was stated that time and motion studies were put to work on a broad front in the textile industry at least during the last decade (that is between the mid-30s and the mid-40s). During this period there had only been a few disputes out in the workplaces. Only in quite exceptional cases had negotiations between the main organizations been necessary. Time and motion studies had not been a "seed of social dissension". In many cases time and motion studies had been requested by the workers themselves. This had even had the effect that there had been no time for method studies because of the workers' demands for control studies.

The textile industry presented special difficulties of a technical character for the time and motion study men engaged in the work. The techniques used were usually largely suited to the conditions applying in the metal trades industry. What was peculiar for the textile industry, which meant that special demands were placed on those managing the time and motion studies, could be covered by the phrases "dependence on material" and "lost time".

With dependence on material was meant that productivity in the textile industry was more directly dependent on the quality of material than in, for example, the metal industry where the even quality of the raw material could be more strictly controlled. This element of uncertainty was connected with lost time: only by prolonged studies could variations in the quality of the material be considered as evened out. If the controlling factor in productivity is the number of yarn breakages in the warp or weft, a short ten-minute study could not be used as a basis either for changes in methods or, still less, for the setting of piece wages. These conditions also, said the author, brought demands for a detailed professional knowledge and knowledge of the branch in those responsible for the management of time and motion studies. It can be noted here how extremely well the line of argument in this particular paragraph lends itself to the view expressed by the workers. Training of time and motion study personnel for the textile industry was looked after by the Textile Institute in Norrköping and Borås, and also by the Institute for the Training of Foremen.[14]

One reason why job studies functioned with so little trouble both in changes in methods and the setting of piece wages the textile industry was, said the author, that there traditionally existed a degree of openness about the studies. There was found paragraph 7 clause 2c, which regulated the carrying out of time and motion studies in the national agreement. It is notable here how the clause had changed character: from being an attempt to prevent time and motion studies it is now regarded as an advantageous regulation of them!

In many cases the national agreement was completed with wage ladders agreed locally, agreements about different levels of earnings for different tasks. These agreements show not least that both parties accepted the scientifically established time and motion study system. No need of further regulation was therefore required.

With regard to the workers' demand for a share in the profits of rationalization the following could be advanced. The gains from rationalization of individual measures could seldom be calculated with sufficient accuracy. Why should only the workers in the department rationalized be rewarded? It would be better if the company rationalized could raise wages to a certain extent generally. The solution lay in wage ladders: if rationalization meant that the work was simplified it was only fair that the workers did not have any direct share in the profits. If on the other hand the work was made more difficult a higher wage was reasonable. Otherwise the question about the profits of rationalization should be taken up in the regular way in connection with the national collective negotiations, which was customary in companies of the textile industry. Thus the experience had been that the workers' attitude to time and motion studies was usually decided by their direct effect on the working wage, especially in those cases where the workers considered the studies were solely utilized for the setting of piece wages. If the profits of rationalization created by time and motion studies were distributed in general increases through negotiations in the national agreement then mistrust by the workers could be forestalled.

Finally, with regard to the organ for collaboration proposed by the Labour Market Committee the Textile Employers Association considered this could only result in further bureaucracy and therefore the association was decidedly against its establishment.[15] This negative stand was also confirmed later by the general council.[16] A year later, when the negotiations in the Labour Market Committee had reached the point of a draft of the collaboration agreement on time and motion studies, this attitude was still maintained by the general council of the Textile Employers Association even though it did not have any objections to the draft agreement presented.[17]

4. Summary and remarks

To recapitulate it can be said that independent statements of policy on the time and motion study problem were not especially frequent on either side in the textile industry. The Textile Employers Association did not raise questions about production unless there was a special reason. When direct questions were put, as by the Rationalization Committee in 1936 or by the Employers Confederation ten years later, the statements made were those dealt with here.

At the first opportunity the right to time and motion studies as obvious necessity was defended. Behind this lay the opposition's attempt to prevent the use of time and motion studies in the setting of piece wages, for example by introducing the provision on production investigations and interpreting it strictly. At the second opportunity was emphasized instead, apart from the importance of being able to use job studies, the concord and spirit of collaboration in which such studies were carried out within the textile industry. In this context the provisions of the agreement about production investigations become something positive, a foundation for this collaboration. Thus on this point the change in the mentality of the employers is seen. It was quite obvious that the workers and their organizations should have insight into and be able to influence the grounds established for the setting of piece wages.

The Textile Workers Association in its turn had greater cause to assume the initiative in statements about time and motion studies, especially with regard to that part concerned with rationalization. During the mid-1930s the arguments used by the union followed a consistent line. Time and motion studies were rejected as a basis for the setting of piece wages and what the employers wanted to use them for otherwise lay outside the field of activity of the organization in the event that they did not cause any difficulty for the workers.

When this position was codified by the Textile Workers Union's Congress the Labour Court had already adopted the stand that the textile workers could not oppose time and motion studies. The attitude of the union during the course of the collective negotiations 1936–1937 also demonstrated the much milder attitude to such studies than the categorical rejection in the statement by Congress. Production control as a way of solving central disputes also became more usual from the mid-1930s. Taken together this suggests that the statement of Congress in 1936 was very rapidly reduced to a rhetoric of decreasing content. The agreement at the middle of the 1940s confirms this impression. The wage ladders almost presumed time and motion studies. During the MYA dispute of 1947 community of view at a high level was confirmed. Here was also to be found the changing attitude of the workers: time and motion studies were recognized.

CHAPTER 14

Piece wages and job studies in the textile agreement 1930–1948

1. Summing up

Here the parties of the textile industry are the subject of attention. Their attempts to adjust to the altered conditions of production brought about by rationalization by means of changes in the national agreement are those initially examined. Their attitudes and decisions with regard to rationalization in general and time and motion studies in particular are also analyzed. Around the core described has been built up a comparatively broadly designed presentation.

The investigation has followed the parties from the time of the greatest conflict in the history of the Swedish textile industry, the great strike of 1931, up to the latter half of the 1940s. During this space of time, it has appeared, large changes took place in the parties' attitudes to the scope of the national agreement, to the regulation of piece wage questions and the significance and role of time and motion studies.

It was the workers who took the offensive in 1931 with demands involving large increases in schedule wages and targets for piece wages. The trade unions should receive influence on piece wage agreements, and the intricate basis of calculation produced by time and motion studies was to be kept out of the textile industry by means of the provision on production investigations.

A more radical section of the Textile Workers Union compelled the union to issue notice of termination of the agreement and the time selected was to appear a bad choice. Not least with the help of the depression the other side, the Swedish Textile Employers Association, was increasingly able to gain the upper hand.

The task of the employers was to hold down the wage level, which was specially accentuated by the approaching industrial crisis, and to avoid such regulation by agreement of conditions associated with production and piece wages as were considered to be reserved exclusively for the decision of management. There belonged the right to utilize time and motion studies and there belonged the right to conclude agreements on piece wages with individual workers.

The long, dramatic collective negotiations of 1931 finally resulted in an agreement that gave something to both sides but tipped the scales in favour of the employers. These succeeded in forcing through their main demand, an immediate lowering of all piece wages above a certain level of earnings. As a quid pro quo was accepted an absolute minimum for the piece wage target; a success for the workers. An aggressive demand for influence on the setting of piece wages and for increased pay that the workers had started with was, however, not met. But they saw as a gain that the provision on production investigation was again found in the new agreement. Nevertheless the development was to show that the hopes of the Textile Workers Union's leaders about the provision were to be disappointed.

During the years immediately following the conflict the workers were rather subdued whilst the employers were more prepared to advance their positions. For various reasons, however, the Textile Employers Association refrained from terminations of the agreement. Up to the middle of the decade the economic situation had so developed that the Textile Workers Union obtained greater scope for action. However, no notice of termination was to be given before 1936.

The collective negotiations leading to the agreement of 1937 were spiced by struggles on many important changes. The provisional piece wage schedules were introduced and relieved the rigidity of the piece wage system feared by the employers. In return the influence of the local trade union organizations was ensured as the right was given to the individual worker to consult his union. Taken together this meant in practice that the fixed schedules, which were approved by the main organisations, gradually disappeared and local agreements on piece wages were concluded between the management and local trade union. On the other hand the ideas about central recommendations for fixed earnings' levels were rejected, levels that during the 1940s were to take the form of so-called wage ladders.

During the later collective negotiations on the agreement during the 1930s the earnings-level was raised by increased schedule wages whilst the procedure for piece wage revision of 1931 was still employed.

The war years during the first half of the 1940s meant for the parties of the textile industry that they were bound to the cost-of-living index agreement concluded by the Employers Confederation and the Trade Union Confederation. A certain overall increase in hourly wages was nevertheless worked out during the period; the economic situation of the textile industry during the war was not at all bad. After 1943, when the effect of the pay freeze was beginning to subside, the workers also demanded higher targets for piece wages. This demand met a cold reception from the employers who, however, were compelled to accept an increasingly rapid wage drift for piece wages and the de facto higher earnings. In order to deal with this (the employers' point of view) and in order to increase influence on the setting of piece wages (the workers' point of view) the parties agreed in 1947 to give up the nonreciprocal relation between hourly pay and piece wages and to normalize in absolute figures the piece wage targets for the different cost-of-living index localities. As early as two years previously provisions had been introduced into the agreement which presumed and gave authority to the local wage ladders that began to appear more frequently. Ladders applicable nationally were discussed. Altogether this implied that piece wage earnings tended to become the subject of central regulation in concord between the main organizations.

Clearly the parties drew closer together in their opinions of which grounds were acceptable in the setting op piece wages. The Textile Workers Union gave up its attitude of total rejection of time and motion studies after suffering defeat in the Labour Court in 1934 and after having been compelled in the 1936/37 negotiations to concede that the provisions on production investigations did not exclude time and motion studies. Instead it was discovered that the provisions on production investigation gave the workers an important possibility of sharing in time and motion studies. It became increasingly usual that the main organizations decided to indicate production control as a way of clearing up disputes about piece wage setting. At the middle of the 1940s both parties made known their approval of time and motion studies as a tool for creating an objective basis for piece wage agreements. Irritation about these studies was negligiable.

2. Remarks

The development summarized above will be linked up with the theoretical assumptions about the parties' actions put forward at the beginning of this presentation. As the development in different areas of agreement is interpreted from the same points of departure and the same conceptions this should create possibilities for a comparison between the metal trades industry and the textile industry.

Regarded over the whole period dealt with it is obvious that the employers were unsuccessful in keeping piece wage questions outside the framework of the agreement. Right at the commencement of the period investigated important sections of the conditions covering piece wages were laid down by agreement, for example the appearance of fixed piece wage schedules and the normalization of piece wage earnings. In 1931 provisions on the grounds for calculation were introduced and in 1936 a threatening central union influence was exchanged for an increased local trade union influence. A certain amount of flexibility was achieved in this way for the locally stronger party.

The boom years with their shortage of manpower, wage drift and tendencies to increased centralization over the whole of the trade union area during the 1940s carried with them a trend against increased central regulation in the piece wage provisions of the textile industry as well. This culminated in the fixed piece wage targets and discussions about wage ladders applicable nationally, constructed on grounds of calculation determined by time and motion studies and approved by both parties.

It is also obvious that the workers had some degree of success in their attempts to increase union influence by putting the weight of the union organizations into their side of the scale. The 1937 agreement especially, brought about an important change that in practice made the local union an accepted party in negotiations about piece wage agreements.

On the other hand the Textile Workers Union failed to keep the more complicated synthetic systems of calculation at a distance. Time and motion studies found a home in the textile industry too. This was of course less detrimental for the workers as their organization, with greater possibilities of the requisite knowledge, had obtained some influence.

The influence of the central organizations also tended, through increase in the volume of piece wages and emphasis of the importance of the fixed schedules, to become increasingly greater. This development was broken off temporarily at the middle of the 1930s but was resumed during the following decade. At that time both organizations seemed to be carried away by the idea of a completely regulated setting of wages with fixed wage steps and "objective" time and motion studies as the foundation. But this tendency was also to be stopped by the demand for local flexibility, this time in the shape of wage drift.

The role of work studies in the relations between the parties may finally be characterized in the following way. When they were introduced by management out in the companies they were a tool in the hands of one party: they were the employers' means of discovering potential economies. The employers saw these studies as a question of planning production in a functional way. The studies came under paragraph 23, they were the private concern of management.

But these economies could affect the earnings-level of the workers. Time and motion studies were then regarded by these as a threat to the balance of the local piece wage agreements. The Textile Workers Union chose to accept their opponents' conditions and to regard time and motion studies as a production question. They maintained therefore that these studies should not influence questions of pay, which were matters for negotiation.

In the long run neither party could maintain its argument. That of the Textile Workers Union was expressly denied by the Labour Court. Time and motion studies undeniably had two purposes of which the one could be considered to arrange production in a functional manner and the second to constitute a basis for the allocation of profits. Neither purpose could be isolated from the other. The threat to the balance of the parties, which the time and motion studies could be regarded as at the beginning, was averted as both parties changed their views on time and motion studies and the question of influence. The workers accepted time and motion studies as a basis of piece wages to the degree that the union's representative was able to refer to them as "objective investigations". The employers agreed that the workers should have the right of joint determination on how the time and motion studies were carried out and they could check the results.

Time and motion studies—the term alone immediately gives the idea of being able to create a system for the objective and just regulation of the apportionment of earnings and a maximum efficient production—also came to be one of the fundamentals of the rapprochement of the parties in the sector covered by the textile agreement.

PART 4

The Central Organizations and the General Agreement on Work studies

CHAPTER 15

Path to the job study agreement 1948

1. Motion on work studies in the Swedish Riksdag

At the reconvened policy conference of the Metal Workers Union in June 1945 to discuss the agreement, when the metal trades strike was approaching its end, Erik Collin, the union's representative, defended the line of the executive board against the communist attacks (cf. pp. 125 et seq.). One of his arguments referred to a decision of the Riksdag on the time and motion study question as justification that the union demands could be constructed in a flexible manner. Behind his argument lies a conception, typical of the times, of the common aims and the complementary arsenal of means of the workers' movement: what could not be obtained by the union could be obtained by political means. This attitude dominated the postwar labour program, which was worked out in the joint cooperation organ of the Swedish Social Democratic Party and the Swedish Trade Union Confederation.[1]

The Riksdag resolution to which Collin referred was the result of two similarly worded motions in both chambers.[2] The mover in the first chambers was the iron worker Sven Boman and in the second chamber the metal workers P. A. Sköldin, Göran Petterson and Sven Hedqvist, together with Adolf Wallentheim, secretary. All of these were social democrats and they had common experiences of the metal industry.

The proposers linked up with the attention the press had given to time and motion studies because of certain incidents. They said that time and motion studies carried with them a tendency to increased psychological and physical pressure on the workers. Additionally there was a risk that earnings from piece wages would be forced down. But these were, it was stressed, the negative consequences of degenerate variants. All Swedish workers understood the absolute necessity of rationalization and the justification of such studies in order to increase production. The last point of view, often repeated, served to mark the fact that they were sensible social democrats, not irresponsible communists, who were talking.

It was obvious, said the proposers, that a great deal of the responsibility for the growth of these unsatisfactory conditions must be considered to fall on the many unsuitable or insufficiently trained time and motion study men who were to be found in industry.

Should then the government interfere in these conditions of industry? Yes, replied the proposers. The ice had been broken for along time. When an unsatisfactory state of affairs existed in industry then the government should institute regulations: industrial welfare legislation, working hours and holidays, and the conciliation institution. Analagous with this

> it must be said that the interest of society should make itself felt so soon as it can reasonably be presumed that time and motion studies are carried on in such a manner that the workforce is damaged by them.[3]

The responsibility of society was so much greater because the private companies had not been made socially responsible in any way.

On the other hand it is more difficult to say what society should do. The motion resulted in a demand for a general investigation. This should consider how the government could help in the better training of time and motion study men, how society could help in creating organizations to guarantee joint influence for the workers and a share in the profits of rationalization and, finally, how the greater spread of time and motion studies would influence public health.[4]

In the last of these demands the proposals linked up with a motion put forward by Fredrik Ström in the Riksdag in 1941. It was also in line with another motion in the second chamber about the organization and research in industrial physiology supported by the state.[5]

The debates in committee had the result that the appropriate committees in both the first and second chambers confirmed the demand set forth in the motions with regard to the training and control of time and motion study men.[6] The problem of effects on public health did not become the subject of any special investigation: this lay within the framework of normal activities of the National Institute of Public Health. Neither was it necessary for society to take up the problems of joint influence and participation. The former belonged to the much greater complex of problems concerning industrial democracy and the latter was a pay question and lay within the field of activities of the organizations. Thus the gist of the recommendations made by the committees became requests for a much more limited investigation than the proposers of the motions had intended.

As a basis for the attitude adopted by the committees also lay, in addition to the motions, statements of opinions from a number of different quarters on the questions submitted to them for consideration. Not unexpectely the Employers Confederation and the Federation of Swedish Industries were completely against the investigation. But what is more surprising is that the National Social Welfare Board also adopted a negative attitude. The Trade Union Confederation (LO) and the Central Organization of Salaried

Employees (TCO) stressed the utility of investigation on the first points but rejected the investigation on questions about pay.[7]

At the beginning of May the matter was taken up in the second chamber whilst the debate and decisions of the first chamber still had to wait another month.[8] The debates in the two chambers of the Riksdag did not add many arguments to those already mentioned above. Those who entered the debate in the second chamber had all been engaged earlier as proposers or as dissentients in the committee. The only newcomer was a communist, a metal worker from Gothenburg. His contribution was marked by personal experiences and lacked the assurance of the social democratic argument about the advantages of rationalization. The fresh argument in the debate was provided by a conservative, one of the dissentients. He reinforced his argument that the question should be left to the care of the organizations with the statement that the Employers Confederation had taken the initiative for discussions with the Trade Union Confederation about the questions under discussion:

As I have said, I do not deny that society has an interest here. But interference from society—about which, I presume, most of us are in agreement—should not be carried out unnecessarily, and the position in this question is such, of which I presume most are aware, that the Federation of Swedish Industries and the Employers Confederation have invited the Trade Union Confederation jointly with both of the organizations just mentioned, to attempt to arrive at a solution of this issue so important to the labour market. What is aimed at here is more accurately, with joint energies to try to create an organ that can draw up the directional lines for time and motion studies. Why should it not be possible to wait and see, one asks oneself, if the parties themselves can in a satisfactory manner arrange the conditions and thus also, perhaps to the degree required, satisfy the societal interests spoken about.

This argument was met by the proposer Sköldin:

It should perhaps not be too much to say that the motion thus confirmed by the dissentients has already had a certain effect. The invitation of the Swedish Employers Confederation to the Swedish Trade Union Confederation concerning the joint consideration of these problems should be possible to interpret as an admission that everything is not as it should be in this area. If the motion and the discussion carried on in connection with it have contributed to this then as the mover I have every cause to feel satisfied.[10]

This is not the place to decide which of the debaters was "right" in his version of the connection between the motion and the beginning of the collaboration between the main organizations. What is important is to note how intimately associated were the political and trade union areas in the eyes of both speakers: initiative in one area may justify the starting or rejection of action in the other.

The majorities in both chambers took the line of the propoers—in the first chamber following a very short debate with well-known arguments—and the Riksdag made the demand to the government for investigation regarding the training and possible authorization of time and study men and the possibility of ensuring the contributory influence of the workers in the carrying out of time and motion studies.[11]

The result, which was never put into practice, must in the light of the continued development be given a special meaning. The demand made to the government had to be regarded rather as a demand for preparatory action than as a call for immediate action. The decision made by the Riksdag put pressure on the organizations, especially on the Employers Confederation.

2. The time and motion study agreement

In another context I have pointed out the conditions for the joint collaboration between the main organizations which occurred during the year immediately following the industrial peace conference in 1928.[12] In this it has been my intention to tone down the importance of the conservative government's introduction of Mondian influence from England and instead to show a domestic line of development over a period of years. Under the auspices of ecclesiastical representatives and later of more practical representatives of the rationalization movement discussions were carried on by representatives of employers and workers. The conditions that these discussions should take place and the foundation for the exchanges at the industrial peace conference was a community of views that developed around rationalization and its need of mutual understanding. This also marks the cooperation between the Employers Confederation and the Trade Union Confederation in the industrial peace delegation and the industrial peace committee set up after the conference. In the report from the delegations both main organizations for the first time made a joint statement about the benefits and conditions of rationalization. After the tragic shorts at Ådalen in May 1931 (cf. p. 54) the Trade Union Confederation withdrew from its cooperation with the Employers Confederation in the newly created Labour Market Committee. This break, however, was only a parenthesis as the development was to show.

In the spring of 1936, following the obvious hints an official investigation, when the parties returned to the negotiation table in order to discuss principal problems in their relations, the committe was revived.[13] Following this the Labour Market Committee continued to be the collaboration organ on questions of principle for the main organizations. This should not be confused with the Labour Market Council, which was set up through the main agreement in 1938 and given certain tasks specified in that agreement.

The Labour Market Committee was the natural forum for the organizations when they wanted to start discussions on a question such as time and motion study. The Employers Confederation's more formal invitation or initiative cannot be dealt with in this presentation but the issue should be dealt with the moment it reached the committee.

At its meeting at the end of August 1945, thus some months after the Riksdag had decided to request an investigation into time and motion studies, there were many important matters up for discussion in the Labour Market Committee.[14] On the proposal of the workers "the problem of collaboration and security" was taken up. This was the beginning of the various discussions that resulted directly in the collaboration agreement on joint works councils a year later.[15] Already at this introductory discussion August Lindberg, chairman of the Trade Union Confederation, stated that it would be natural for these contact organs now considered for companies to handle rationalization questions.

Time and motion studies were included as a special item on the agenda of the Committee. Gösta Ingelman reported on time and motion studies in the metal trades industry and about the training of time and motion study men that since 1944 had been carried on under the management of the Metal Trades Employers Association. Erik Collin presented the complaints that the workers wished to put forward and proposed definite measures: the removal of unsuitable job study men, better training aimed at greater impartiality, greater influence for the workers by means of provisions in the agreement and increased information.

A concrete program was also presented by Gösta Ekelöf, who regretted that training of job study men was spread among so many hands and stressed that the courses given at the Institute for the Training of Foremen (ALI), of which Ekelöf was the manager, aimed at being factual and impartial. To the description of this institute activities was added a sketch of the arrangement of courses that could be approved by both parties. Ekelöf drafted guiding lines for an organization for the control of training and dealing with other time and motion study questions that had borrowed the main lines from the forms of collaboration in apprenticeship questions, with committees in each area of agreement and a central council.[16] It was proposed that a central time and motion study council should be set up. This should be

> a forum for time and motion study questions in industry, to that degrees these may be said to be similar for the whole industry and not of a character that they belonged to the areas of negotiations of the respective employer organizations.[17]

Ekelöf further described the tasks of the proposed council. In general it should follow development in the area; it should draw up the qualifications required for job study men and prepare instructions for training and possibly issue certificates of competence; it should dimension the training at ALI as required and inspect the training there; finally it should influence the individual companies to maintain in their internal training the qualification requirements laid down and not to employ work study personnel without such competence.

The job study committees should carry out their activities by branches, said Ekelöf, with the task of seeing that the directions of the council were followed and supply the council with the necessary empirical grounds for decision. Both the council and committees should be constituted from both parties. Finally, Ekelöf proposed that a research and investigation department should be linked to ALI with a least one full time principal.

In Ekelöf's program can be observed the association with the newly-signed metal trades agreement with regard to the time and motion study council. The main features were also to recur in the time and motion study agreement created later. On the other hand Ekelöf was not exactly successful in his obvious attempt to strengthen ALI in its debatable position between the two main parties.[18]

On behalf of the Employers Federation Frithiof Söderbäck said that Ekelöf's program could form a basis for the continued discussion with the exception of the point on the investigation department. August Lindberg adopted a wait-and-see policy and proposed that the committee should return to this question later, and this was approved.[19]

At a later meeting of the committee a delegation was appointed to investigate the question jointly.[20] In addition to a general discussion about time and motion studies, not reported in detail in the minutes, this problem also recurred in connection with the continued discussions about industrial democracy.

Arnold Sölvén, a solicitor of the Trade Union Confederation, who functioned as one of the committee's secretaries, put the view of the workers' side in a memorandum.[21] In this reappeared the well-known assocation between responsibility for the efficiency of trade on one hand and joint determination on the other.[22] The tones from the social democratic arguments in the early proposal of 1923 on joint works councils were heard still more clearly.[23]

The argument, to a great extent to be seen again in the joint works councils' agreement, was thus securely based on the tradition of the Trade Union Confederation to encourage efficiency. Sölvén drew up a number of tasks for the organ of industrial democracy:

> As the principle and in practice first of the *tasks* it is considered should rest on the organ of industrial democracy must be considered the task of furnishing the employees with increased insight *into the techniques and functioning of production* and to stimulate and utilize their interest for the more active collaboration in the attempts to achieve increased and improved production results . . .
>
> The second important task is to supply the workers with insight into the *economic conditions and results of production.*[24]

Somewhat later Sölvén took up the question of time and motion studies directly, but here the principal difficulty was brought to the fore, that was inherent in all proposals about industrial democracy and which referred to the conditions between joint works councils and trade union organizations:

> Whilst the production and welfare questions now stated naturally fall within the frame work of the activities of the organs now concerned, it is a little more doubtful whether these organs should also deal with the question of *time and motion studies*. It should be generally basic that the organs should not have the right to deal with questions on the regulation of the conditions of work. ... Actual disputes concerning the carrying out of time and motion studies in special cases should, like disputes about the setting of piece wages on the basis of such studies, be dealt with by the procedure now applicable. On the other hand there seems to be no objection that questions about the methods of time and motion studies, as on the whole methods for wage setting, are made the subject of discussion about principles by the new organs.[25]

At the five meetings of the committee which followed the discussions were to be dominated by the question of joint works councils. The time and motions study question was put aside. The delegates appointed in November 1945 had already prepared a memorandum by February 1946 that, however, was not presented to the committee before the end of June. It was at this meeting that the committee agreed upon a final draft agreement on joint works councils. Nor at this meeting did the committee have any discussions about the time and motion study question.[26]

The memorandum from the delegates confirmed initially that efficient production was of just as much interest to the workers as to the employers. The workers were aware of the value of time and motion studies but indicated sources of irritation. Method studies and the improvements brought about them in the apparatus of production were too often absent. Similarly with studies about lost time, which led to faulty lost time bonuses calculated to a standard pattern. Time and motion studies were often considered only as a means to lower the piece wage. The job study men were often incompetent. The workers did not take part in the planning or in the profit-sharing.[27] As can be seen, the list contained nothing new.

The delegation suggested that the committee should make statements of a comparatively general character on the various items in the list of causes of complaint. Even more generally worded was the argument put forward in connection with the question of the sharing of rationalization's profits. It was definitely proposed, in accordance with Ekelöf's program, that a central work study council should be set up in every area of agreement. It should comprise an equal number of members from each side and its tasks similarly formulated to what had been laid down for comparative organs in the metal trades agreement. The idea of local time and motion study councils was rejected, however, with reference to the joint works councils in the making and the already existing safety committees. The Employers Confederation and the Trade Union Confederation should set up a joint time and motion study council with the task of acting in an advisory capacity at courses arranged by branches of the organizations concerned and also to deal with time and motion study disputes that arose.[28]

Not until April 1947 did the Labour Market Committee return to the time and motion study question. A draft agreement was presented by the Trade Union Confederation. The employers then stated that the actual question involved were mainly covered by the agreement on joint works councils and that the committee would therefore be satisfied with a recommendation that could form a basis for dealing with the time and motion study question in joint work councils and the Labour Market Council.[29] From what has already been said it seems that the employers' attitude was not without justification. Sölvén had already advanced this idea. But the workers were out after an agreement, not a recommendation that ran the risk of being much too weak. So the time had arrived to play the card that had been held for about two years. The minute continued: "It is recollected that the attention of the government has already been drawn to the time and motion study question. It was considered by the workers' side that only an agreement could be regarded as a real alternative to legislation."[30] With this very slightly veiled threat the question was shelved. It was no longer a case of the creation or not of a special organization for collaboration on time and motion studies but what it was to look like.

In November the Employers Confederation presented its counterproposal for the agreement. The other side was unable to reply immediately but it was thought that the proposal would probably be acceptable with only smaller adjustments. With regard to the introductory general explanation, the differences were greater. The question was once again shelved.[31]

At the meeting at the beginning of May, 1948, each side presented a reworked proposal. On the whole these agreed but differed on some points. During the discussions the committee succeeded in agreeing on a joint proposal.[32] This was approved by the parties without alteration at a later meeting with the committee.[33]

The proposal for a time and motion study agreement consisted of two parts: a general statement of reasons and a draft agreement. The general motivation, where most of the differences between the parties' proposals were to be found, commenced with a confirmation of the importance of increasing efficiency and the role of time and motion studies. Then the possibilities of local collaboration were confirmed. A clause on the carrying out of time and motion studies followed where the principles for the setting of piece wages by means of time and motion studies were touched on. Finally the required qualifications for study men were dealt with. The draft contained a clause about time and motion study committees at the branch level and one about the central time and motion study council.

The differences between the parties as they appeared in the proposal to the meeting in May 1948 may be delineated in the following manner. As a basis for the drafting lay the delegates' memorandum of February 1946. In their motivations for the agreement both parties tried to tie down the other to important points in the arguments to be found in the memorandum. In the Trade Union Confederation's proposal the clause on the prosecution of time and motion studies was a great deal more extensive than it was finally. The workers tried to limit the drive of the employers and their job study men in the field which had earlier been their unlimited domain. In line with this the proposal from the Trade Union Confederation aimed at defining exactly the role of the job analysts in collective negotiations. In principle they should only be advisory to the employers on questions of fact, not their representative in the negotiations. On the former of these divergencies there is no trace of the proposal of the workers in the final agreement, whilst with regard to the latter the Trade Union Confederation's wording is stated as a rule, although at the same time it is admitted that in certain cases the study men could also function as negotiators.

From their side of the fence the employers tried to tie the Trade Union Confederation to the line taken by the memorandum on the issue of the sharing of profits and succeeded to a certain extent. By pointing out in the introductory paragraph that the employees, as every member of society, benefited from rationalization some of the energy was dissipated in arguments about the workers' share in the profits of rationalization. Another point where the Employers Confederation managed to get their proposal accepted for drafting was the reminder about the right of study men to support and loyalty from both sides.

An interesting difference lay finally in the statements made on the existing possibilities of local collaboration. The employers, whose proposal was followed, first named the trade union organization and then the new joint works councils. The workers had the reverse order. However, it is impossible to determine if this difference should be interpreted as dissimilar views on which questions should presumably be activated or as different evaluations by the different organs themselves.

A similar difference of opinion applied in the construction of the provisions in the agreement concerning the tasks of the time and motion study council. The employers wanted to include in the agreement that the council should promote collaboration and usefulness in job studies. The Trade Union Confederation wanted to change this order: to work for usefulness to stand as the first task, and to work for collaboration as number two. The final result was a compromise with a bias towards the employers. To summarize, it is obvious that in the final drafting the employers managed to get their proposals through on more points than the workers. It is just as obvious however that the agreement was the fruits of the attempts by the workers' movement to gain influence over important factors in production and changes in these.

3. Summary

In line with the basic considerations behind the postwar program of the labour movement, the united movement tried to obtain influence over rationalization and time and motion studies by utilizing a comibination of union and political measures. Simply by raising the question in the Riksdag the workers put pressure on the Employers Confederation in order to bring about negotiations. The decision of the Riksdag to request a governmental investigation became a continued support for the Trade Union Confederation. In a situation during these negotiations where the employers seemed to want to apply the brakes this latent threat was utilized as an argument. Once the agreement was in existence the governmental investigation was not completed.

The birth of the time and motion study agreement resulted from a trend in affairs running parallel with and woven into that which led up to the agreement on joint works councils. It seems even more correct to regard both of the agreements as the fruits of the same development and striving. The time and motion study agreement was an extention of the joint works council agreement, brought about by a special need of coordination mainly in the training area. Both agreements had been well established in the Trade Union Confederation for some decades in its confirmed line on production policy. The general plan of the time and motion study agreement provided expression for what was just and reasonable in the workers' demand for joint determination in rationalization measures at the practical level, which had been formed during the work of the industrial peace delegation twenty years earlier. To a corresponding degree the agreement was a setback for those among the employers who still regarded time and motion studies as the private concern of company management. At the same time, the investigation has shown that the construction of the agreement was marked in large sections by proposals made by the employers on organization and wording. Solicitude about production, the basic element of the ideas of rationalization, was then manifested as a workable foundation in the striving for compromise in industry.

Within the embrace of this compromise were crushed the original attempts of the representatives of rationalization to assume an intermediate position. The attempt by Gösta Ekelöf to establish the Institute for the Training of Foremen as an independent institution between the Swedish Employers Confederation and the Trade Union Confederation achieved no success.

PART 5

Conclusions

CHAPTER 16

In retrospect

In this final part of the book the questions originally raised will be discussed in the light of the experience gained from the sectors covered by the metal trades agreement and the textile trade agreement.

1. Utilization of job studies

It seems reasonable to consider first to what degree time and motion studies, here understood as a significant expression of the new technology of production that made its break through during the interwar period, actually came to be used in the branches of industry investigated.

It has not been possible to present an exact picture stemming from the central material utilized in this study. But comparatively great clarity can be obtained for the metal trades industry. Among the member companies of the Metal Trades Employers Association 10 plants employing almost 12 000 workers used time and motion studies in 1920 at the beginning of the period investigated. A decade later this number had trebled and by the end of the interwar years time and motion studies were used in over 800 plants. More than 60 000 metal trades workers were affected.

It is impossible to make any comparison with the textile industry on this point as no comparative internal material has been discovered. From the investigations of the rationalization committee it is known, however, that time and motion studies were used at the middle of the 1930s in all branches of the textile industry, especially in the largest companies.

The last observation gives rise to reflections about the possible connection between the structure of industry and the utilization of time and motion studies. There seems to be an obvious dependence between these studies and size of company. Possibly there is also a relation between time and motion studies and a bias towards export. Both within the metal trades and textile industries it was the large companies who had the resources for and felt the need of introducing job study techniques. In the wake of these pioneers followed other larger and medium companies. As I have been able to indicate in another context, this phase of development was supported by external auxiliary agencies, consultants or expertise linked to the branch organizations.[1] The medium sized companies were unable within their own organizations to devote special resources to develop and apply the techniques of time and motion studies. It should be noted that to a large extent it was during this phase and about this category that the disturbances and irritation arose with them.

It is possible to formulate a more general conclusion in the following way. An industrial structure with fewer and larger companies should constitute a better basis for new ideas on the technologies of production and organization to make a breakthrough than in a structure consisting of smaller and more units. A larger number of smaller companies also increases the risk that the slow penetration of innovative ideas is accompanied by irritation and dissatisfaction about faulty applications.

The brittle information on the spread of time and motion study technology has not shown any noticeable coincident variation with the development of the economic situation. In another context I have shown that the crisis at the beginning of the 1930s triggered an activity aimed at disseminating the ideas of rationalization, including time and motion studies, but the information about the actual conditions does not show that these efforts produced any obvious immediate results during the years of crisis.[2] On the other hand it is quite clear that the dissemination of the new techniques proceeded considerably faster during the 1930s than during the previous decade and that this should also be carried back still further to the fact that agencies were also created and extended for the dissemination of ideas during the years of depression. It should also be observed that the material selected here does not suggest that the years of depression at the beginning of the 1920s and the beginning of the 1930s hampered the spread to any noticeable degree.

2. The piece wage system and conditions for the production process

Job studies were the expression of the new techniques for the planning and organizing of work. Modern ideas or simply the thought of them were received in different ways by the employers and the workers. For the employers "scientific management" appeared as the promising alternative worth trying in order to make manufacturing more economic. For the workers such studies were a tool in the hands of management to reduce earnings and increase the speed of work. Common for both parties was the idea of job studies, the new technique, as a new factor in one of the scales balancing their relationship.

Once time and motion studies were introduced they changed the conditions between the parties of the labour market. It was up to the organizations concerned to react in such a manner that their own benefits through, or in spite of, changes in technique became as large as possible whilst at the same time the disadvantages, the advantages of the other side, were thrust aside.

The component initially affected in the relations between the parties was the agreements about piece wages. This particular component was obviously extremely vital. Work on piece wages gave higher earnings than the toil of hourly wages. When they had once accepted the principles of the piece wage system, the Metal Workers Union and the Textile Workers Union tried to increase the piece wage volume in their fields. It was up to the trade union organizations to safeguard the interests of their members in the rising number of piece wage agreements.

The employers' organizations, especially the Metal Trades Employers Association, opposed this. They proceeded from the idea that piece wage agreements were free settlements between management and the workers who had to decide about the piece wage and work at that wage. Nor right of influence from the trade union organizations was admitted. Every regulation of the use of piece wages was therefore to be regarded as a gain for the workers.

The endeavours to ensure the interests of the members in the setting of piece wages was of course just as fundamental both in the Metal Workers Union and the Textile Workers Union. But here the similarity ceases. The grounds for action varied considerably in the different areas covered by agreement.

Textile production was uniform compared to that in the plants of the metal workers. The suboperations in the process were the same; the path of the raw material to the finished product was determined by machines whose character only changed at intervals during the period concerned. Automation increased at a blow the production capacity of machines but did not basically change the relation between men, raw materials and machines. The capacity of the machines set the limits and norms for the production capacity. Within these frameworks the quality and characteristics of the raw material were decisive in how well production proceeded. The need for new piece wages was dependent on deliveries of raw materials and naturally by the design of the intended product. But the series manufactured were often long, which served to keep the number of piece wages down, the number being the cause of the struggles and the dissatisfaction.

The position was different in the metal trades industry. Mixed manufacture and changing products in short series continuously demanded the setting of new piece wages. Annually millions of piece wages were set in the plants.[3] The characteristics of the machines and raw materials were stable and well-known. The most variable factor in piece wage conditions in this industry was the human contribution of labour. The link between productivity and the possibilities to build up the piece wage on the one hand, and the speed of work, experience and skill of the workers on the other was much more direct than in the textile industry where the accent was on the process. The relation between piece wage earnings and intensity of work was nakedly obvious in the daily struggle between management and worker.

Thus the production processes created changeable conditions for the behaviour of the parties in piece wage questions. It was therefore in the nature of the problem that the struggle about piece wages was considerably tougher in the area of the metal trades agreement than in that of the textile agreement. The physical conditions of production can be regarded as the basic explanation of the differences between the branches of industry to be dealt with in the following. The aspects dealt with specially are: the question of influence over the setting of piece wages, the question of how earnings on piece wages were to be ensured, and the question of how time and motion studies were generally received by the parties and to what degree these studies aroused irritation or other antagonisms.

3. Relationship to rationalization

If initially the general attitude and reactions of the organizations to time and motion studies are considered it is clear that both employer organizations first took an interest in them comparatively late. Rationalization was essentially something for the individual company rather than for the branch organizations or other agencies for the stimulation of technical training and development.

When these organizations or agencies approached employer interests too closely, that is to say when in a more obvious way they touched on wage questions, then the employers' organizations reacted. This applied especially to the Metal Trades Employers Association during the first part of the 1940s when the settlement occurred with the Institute for the Training of Foremen and the representatives of the view that maintained the role of rationalization men as impartial expertise. It has not been possible to observe any marked comparative reaction from the Textile Employers Association, but the basic attitude was the same: time and motion studies should be used and did not need to be made the object of any external interference.

Throughout, the Metal Trades Employers Association drove a harder line in piece wage and time and motion study questions than the Textile Employers Association. Here the question concernes part of a general policy; relations between the parties were considered to be considerably better in the textile industry ("the textile spirit") than in the metal trades industry where personal and political differences played a greater part. At the bottom also lay the demands of the physical process of production: the setting of piece wages and time and motion studies were a more vital matter for the managements of the metal trades' plants than for the textile factories.

So it was the organizations of the workers that pressed the regulation of piece wage setting in the national agreements and who brought up the matter of time and motion studies during the negotiations. This took place about 1930, some years earlier in the metal trades' negotiations. So far the conditions were similar. But then the paths diverged. Both unions really reacted quite differently to the challenge of time and motion studies.

For a long time the Metal Workers Union had accepted the piece wage system and with certain reservations were committed to the idea of rationalization. Therefore they were in principle unable not to accept time and motion studies. Instead of opposing time and motion studies as a whole the union attacked its objectionable variants and its misuse. At the same time the management of the union understood towards the end of the 1930s that the union's representatives had to be trained in the methods of time and motion studies. In the Metal Workers Union it was considered that the union representatives had to command the new technique *just as well* as the employers.

The Textile Workers Union followed a completely different path. The union had certainly also accepted piece wages, but by trying to get the simple statistical investigation of production accepted as the sole basis for piece wages the union tried to keep time and motion studies at a safe distance from the textile industry. This was unsuccessful in practice but during the 1930s the union still acted as though job studies did not exist. Its leaders refused to include information about job studies in the union program on the grounds that this could be interpreted as indirect approval of the idea behind the studies. It was to become clear that this line was impossible and by the middle of the 1940s the Textile Workers Union had in practice clearly accepted job studies as the basis for piece wages.

4. Influence on setting of piece wage rates

Assuming this essential basic attitude to job studies attention will be directed to how these studies influenced the way the parties maintained their own interests when these concerned influence on the setting of piece wages and ensuring the earnings-level.

Influence on the setting of wages was obviously an aim in itself for the trade union organizations. It was also a means of ensuring the earnings level during periods when the economic situation depressed wages and a way of guaranteeing the workers a share in increasing profits when times were better.

With regard to hourly pay, the influence of the organizations was conceded at the beginning of the 1920s, the starting point of this investigation. But in the case of piece wages the influence of the union was far from obvious. The union had some control over the fixed piece wages included in piece wage schedules, usually through a local or central approval. But the remainder, or the free piece wages, were more difficult. For the workers the problem was that it was either a case of getting a greater proportion of the piece wages fixed or of obtaining direct influence on the setting of piece wages when free piece wages were involved.

For the employers piece wages were seen as an area where the free agreements between management and worker should be predominant. It was of course part of the system that piece wage earnings should exceed the minimum level laid down in a schedule in the national agreement. Piece wage agreements thus interfered with the obligations to the trade union organizations.

In line with this attitude time and motions studies were introduced as an exclusive tool of the employer in method development and the setting of piece wages. The demand to lower production costs, the demand to rationalize, led the employers increasingly to utilize the recipe that an increasing number of rationalization experts recommended. The old "foreman piece wage" based on experience untried by experiment was replaced by a system of piece wage calculation that made claim to be objective, rational and scientific.

For the workers these new methods became a local source of irritation. As could be expected, dissatisfaction found expression at the central level, earlier in the Metal Workers Union than in the Textile Workers Union. From the end of the 1920s there were repeated complaints about time and motion studies at the policy conferences of the Metal Workers Union. The leadership of the union reshaped these complaints into demands for influence on the way in which time and motion studies were carried out. It was the workers who brought up the question of time and motion studies for negotiation.

The Metal Trades Employers Association looked after its right to independent time and motion studies just as it protected the principles of paragraph 23. Concurrently the use of time and motion studies increased. With the precedent created by the judgment of the Labour Court in 1932 followed the demand that the employer should be able to show that newly-set piece wages would not imply a decrease in earnings. It should be possible, let alone with a more intensive contribution of labour, to earn just as much even on piece wages of a lower nominal value.

The workers' breakthrough into the area of the employers in the metal trades agreement went slowly and by stages.

The agreement of 1936 gave the individual worker the right to be informed of the basis for piece wages. Two years later this right to information was extended to apply also to the individual time and motion study report. The central organizations received the same rights in those cases where a dispute on piece wages arose. In 1944 the local union obtained the right to see all time and motion study material in every case and to request control studies and to take the initiative for new job studies. In 1947 minutes were included in the agreement which aimed at compelling local settlements on money factors. In the following year the time was ripe for centrally agreed targets for the setting of piece wages in the metal trades industry.

The implication of these successive changes in the piece wage provisions of the metal trades agreement was first that the individual workers, followed by their central and finally local organizations, had gained influence in an area where earlier the employers had been solely in command.

For the Metal Workers Union this extension of influence over time and motion studies was a long drawn out and laborious struggle. It was easier for the Textile Workers Union. The provisions on production investigations introduced into the national textile agreement as early as 1931, provided the textile workers at a single blow with much of what the metal workers first achieved after hard tussles over ten years. What was paradoxical in the situation was that the provisions were originally intended on the part of the Textile Workers Union as a bar to job studies. As such the provisions were unsuccessful as the judgement of the Labour Court showed in the Rydboholm case. But this gave the textile workers rights in job studies that were much more far-reaching than those the metal workers had.

One explanation of this unexpected success in the long-term for the Textile Workers Union lies in the circumstances surrounding the genesis of the 1931 agreement. At that time the Textile Employers Association had one aim that overshadowed all others: an immediate and large decrease in the wage costs paid in piece wages. In comparison with this the compromises in other parts of the agreement carried less weight, including the provisions on how production investigations should be carried out. On this point the employers were satisfied with the guarantees that job studies themselves should not be interfered with by the new provisions. It looks as though the employers in the textile industry had never investigated the real implications of the proposed agreement.

The actual conditions of production and such momentary circumstances as the last mentioned were factors that may explain the difference between the metal trades industry and the textile industry. But common to both areas and on a general level there are other explanatory factors as to why the workers were able to make this entry into their opponents' area. It is, seen with the system of negotiations then current, the increasing complexity in the setting of piece wages brought about by job studies, and it is the claims for objectivity raised by job studies which made the increased influence of the workers both necessary and possible.

Advocates of rationalization theories ever since Taylor had stressed that rationalization required collaboration. This association between joint determination and rationalization was also made by the central organizations, the Employers Confederation and the Trade Union Confederation, for example in conjuction with the report of the industrial peace delegation in 1930. A central element in the line of thought was that modern rationalization measures were taken on the basis of scientific considerations.

The tempting arguments for job studies were the same. Job studies would produce a basis for the setting of piece wages that was "objective", which was not caused by differing interests, which was rationally confirmed. The "objective", the scientific, could not in the long term be regarded as the sole interest of either party. The right of both parties to be informed about and to utilize time and motion studies followed logically from the consideration of time and motion study techniques as an objective, scientific method. An employer could not offer what appeared to be a low piece wage based on job studies and argue in favour of its scientifically proved correctness without granting his opposite party the right to be informed of these scientific investigations.

When the individual workers received this right of information it then followed from the negotiation procedure that the main organization also was granted the right if disputes arose on a piece wage setting derived from a job study.

The increasing complexity in the calculation of piece wages found its expression, for example, in the fact that the organizations on both sides considered they were compelled to establish special appointments or groups to deal specially with job study techniques. The chances of the individual worker to make himself heard in piece wage negotiations was limited in practice to the right to accept or refuse a piece wage instead of assessing the grounds on which it was based.

The number of central disputes requiring "studies" increased. The right of the main organization to influence was utilized increasingly because it was for the workers the only instance where both had the right to be informed about job studies and the expertise to assess them.

At the beginning of the 1940s this development tended to swamp the resources of the central organs of both parties. In this situation the right to information and to take the initiative was moved downwards to the local trade union organizations who quickly also received the right to collaborate directly in time and motion studies. For practical reasons management in reality came to negotiate about piece wages with the local trade union organization, which had the necessary skill at a decentralized level. In this way time and motion studies, the expression of a new technology of production and production organization, came via an internal logic to break the opposition of the employers to trade union negotiations in questions about piece wages.

5. Assurance of earnings-level

To ensure the workers an influence in the setting of piece wages through central provisions in the agreement of course aimed at being able to maintain the workers' interests in questions relating to pay. In both the metal trades agreement and the textile agreement there were also attempts at a more direct regulation of the earnings level. It must be mentioned that what the workers' side was more interested in, with regard to the central provisions, was to stress the minimum levels. This therefore became especially pressing during the slump at the beginning of the 1930s. Conditions were otherwise during the following decade. There was then a shortage of labour and this resulted in a periodically powerful wage drift especially for piece wages. In this situation it was not the workers' organizations that were most interested in provisions of national applicability. Instead it was the employers' organizations that wanted to have provisions on a piece wage ceiling, not least to keep the peace in their own camp.

A comparison also shows that when it involved ensuring piece wage earnings, the textile agreement was more favourable than that of the metal workers. From the beginning of the 1920s the textile workers had had the stated piece wage target in their national agreement. Piece wages were to provide an excess of on average 15% above the hourly pay, from 1931 at least 15%. Certainly the target was vague and certainly the union tried all the time to increase the stated guarantee whilst the provision was queried periodically by the employers, but nevertheless there was a central stipulation on the minimum values of piece wages.

In the metal trades agreement there was no comparative guarantee for piece wage earnings. During the particular period concerned here the Metal Workers Union was compelled to try different approaches in order to achieve security of earnings for those of their workers who received piece wages.

In this agreement a difference was made between fixed and other piece wages. For the fixed piece wages in the schedules there was a general earnings-guarantee in what was called minute 1. Under the chairmanship of Fritjof Ekman the Metal Workers Union tried to do away with the difference between the different kinds of piece wages so that all such wages could be brought in under this minute. These tactics were unsuccessful. With the help of the Labour Court the Metal Trades Employers Association was able to maintain the traditional separation. But two years later the security of piece wages on the fixed schedules was stressed by the court. The earnings-guarantee referred to the earnings-level of the individual workers.

The trade unions then tried another approach with an attempt to minimize the number of other piece wages so that in practice all of these should be included in schedules, be considered as fixed and so secured with regard to earnings. The most important aspect of this attempt, however, became the question of the influence exerted by the union organizations, which was dealt with above. The Metal Workers Union was thus never able to obtain central targets in the setting of piece wages by its own efforts. It was not until some two years into the 1940s, when the new situation on the labout market changed the employers ideas, that central piece wage targets, a general guarantee of piece wages, could be realized. The question of stabilization of targets was brought up by the use of time and motions studies.

On this point there is again some reason to consider time and motion studies. It was because of the general use of these in the mechanical metal trades industry that the stabilization of piece wage targets became significant. The fixed cash factors provided standardizing values for piece wage earnings under normal conditions. Job studies were presumed to be the tool with which it should be possible to transform the actual conditions of changing production and compare these with the norms. Development in the textile industry that had resulted in the construction with wage ladders was analogous. For the actual evaluation, that is the grading of the various tasks onto the different steps of ladders, no job studies were required of course. But as a connecting link between a diversified and detailed reality on the one hand and the theoretical norm on the other, time and motion studies were taken for granted. In this manner these studies became the point of departure for the setting of central norms that could offer the earnings guarantee and the piece wage ceiling that the organizations had striven to attain time and again.

6. Trade union opposition

The picture of the development in both fields of agreement now apparent is interesting and provides food for thought. The Textile Workers Union actively opposed the new procedures for the setting of wages and organization of the work. The Metal Workers Union accepted them with certain reservations. Nevertheless it was the Textile Workers Union, paradoxically enough through measures in another direction, which ensured its members both a real influence on the setting of piece wages and a guarantee of earnings earlier than its brothers in the metal trades industry. On these particular points the textile agreement was more advanced than the metal trades agreement.

Because of the design of production conditions the Metal Trades Employers Association fought for free piece wages more intensively and aggressively than the Textile Employers Associations. Nevertheless the employers of the metal trades, during the latter part of this investigation, were compelled in this matter to admit the right of the workers and the trade union organizations, which the Textile Employers' Association had accepted long before.

This impression is confirmed if the disputes about time and motion studies registered at the central level are considered. Certainly it is obvious that the level of investigation here is a source of misunderstanding. The central material only shows the local dissensions that led to central disputes. Obviously there may have been a great deal of annoyance out in the factories and workshops and there may have been many disputes that were dealt with by means of local negotiations, perhaps guided by precedents from earlier central agreements. Bearing this limitation in mind it can be said that the irritation surrounding time and motion studies seems to have been a great deal larger in the metal trades industry than the textile industry. This may to a certain extent be explained by the fact that the number of piece wages subjected to job studies, and thus reasons for conflict, was a great deal larger in the metal trades. But it is also reasonable to regard the more favourable textile agreement, which guaranteed a minimum excess on all piece wages and also the right to influence, as a reason that there were fewer conflicts in the textile industry. One question that arises in this context is how the internal conditions in the trade union organizations were at that time.

Centrally in the Trade Union Confederation ideas on rationalization became the subject of differences of opinion between the social democratic leaders and the majority on the one hand and the communist opposition on the other. It is also clear that in the Metal Trades Union the position on rationalization and job studies was one of the points at issue on which the representatives for more radical lines, communist or unaffiliated unionists showed up in sharp relief against the reformist management of the union. The opposition in the Metal Workers Union was comparatively strong and job studies played a marked role in the union program put forward, for example, in connection with the united front movement at the end of the 1920s and again in the metal trades conflict fifteen years later. As a matter of fact it was in the arguments with this internal opposition that the leaders of the union were compelled to make known their stand on rationalization and job studies. So far as it is possible to see this more aggressive radicalism in the Metal Workers Union raised the temperature of the local disagreements with the employers whilst at the same time it weakened the position of the leaders of the union when they were carrying on negotiations. They were compelled at one and the same time to defend attempts to obtain greater efficiency within the industry against the censures of the opposition, while they were pressing demands on the employers as conditions for accepting the same rationalization.

The Textile Workers Union was able to go ahead with a greater degree of composure and consistency. Of course there was an opposition here too, not least in the conflict of 1931, but its strength was relatively speaking considerably less. Here lies also one of the contributory reasons for the textile spirit mentioned above, the good feeling that existed between the parties of the industry.

7. Professionalization

It has already been said above that time and motion studies and the greater complexity stemming from them in the setting of wages and methods development smoothed the path for an influence from the trade union organization on piece wage questions. Through this the setting of piece wages and job studies came to a higher degree than earlier to be regarded as a matter for both sides. The most obvious expression of this was the creation of the collaboration agreement of 1948 on time and motion studies and the time and motion study council set up through this.

Both of the areas of agreement investigated here were so advanced in their development in job study questions that already within the frameworks of the national agreements methods were given on how disputes about time and motion studies should be dealt with. Assuredly this did not imply that the field covered by time and motion studies was so free of sources of discord; the time and motion (MTM) troubles of the early 1950s were just around the corner. But the procedures for joint collaboration of the parties were laid down in the collective agreements, therefore the central job study agreement really added nothing new to the areas investigated here.

The investigation carried out has also shown that at the beginning time and motion studies resulted in increased influence of the central agencies of the trade union organizations. When the number of cases threatened to overload the available resources of the central bodies concerned, a shift to the lower organizational levels of the ability to deal with such cases took place on both sides. In reality the piece wage agreements at job studies came to be made by the supervisors and a representative of the local union.

This representative system was in itself a manifestation of the tendency to the subdivision of work, a characteristic result of rationalization. Just as planning of the work was separated from its performance, questions of payment were also removed from the direct decision of the workers.

Planning in a rationalized organization was removed from the physical work and thus decreased the actual and potential possibilities of the worker to control his own situation. In the same way the complicated formulae of time and motion studies tore the setting of piece wages from the hands of those who did the work. Also this decreased the chances of the individual worker for responsibility and to take in the situation. In both cases is found dilution of the potential content of an occupational role.

In order to avoid misunderstandings on this point it is stressed that this negative state of affairs obviously had, to a certain extent, a completely dominating, positive side. Because of the increasingly important part played by the trade unions the workers were able to get a bigger slice of the cake in the same way that rationalization, by setting up central planning agencies, had also meant larger cakes being baked in which to share.

Time and motion studies were demanding complicated ways of calculating earnings. They therefore came to stimulate a higher degree of expertise among the few but at the same time they perhaps deprived most of the chances of independent judgement. In the investigation it has been possible to follow how the representatives of the workers in the metal trades requested the help of the union at every policy conference on the agreement and at every congress from the end of the 1920s: about information, study guidance, and especially representatives. From 1938 the time and motion study committee of the Metal Workers Union led the demands that made recommendations to the executive board for a powerful investment in information and training. Some years earlier the Metal Trades Employers Association had also felt the need of special qualifications in the area of time and motion studies and at the beginning of the 1940s followed this up with a special committee among whose tasks was the regional training in time and motions study questions.

In my earlier investigations of the rationalization movement it was maintained that a core of rationalization experts was created and professionalized in the 1930s and 1940s. In this study the reaction of the trade union parties can be described. The Metal Trades Employers Association accepted the demands of professionalization but wanted to seize and control them for their own purposes. The Metal Workers Union also admitted the fundamental justification in the idea of rationalization work as a profession. The Employers Association accepted the argument that irritation with time and motion studies often depended on the insufficient training of job study men. Member of the union also lay behind the motion in the Riksdag about the public training and authorization of time and motion study men in 1945. The aim of the Metal Workers Union in the time and motion study question was that the workers, the central and later the local negotiators, should command the techniques of time and motion study just as well as the employers, that is to say in this respect to acquire the skills on which rested the demand for professionalization of the rationalization men.

Within the textile industry, as has appeared, the result of the development became the same but was not achieved in the same way as the Metal Workers Union. The Textile Workers Union rejected time and motion studies as the basis for piece wages not because in themselves they were erroneous but because they were complicated and difficult for the workers to master. These studies threatened consequences that the textile union was not prepared to accept: professionalization, the gap between those who were competent and those whose competence had been taken away, was what the textile union would not accept. But to set up a barrier against development towards a more complicated society was also doomed to failure. The textile workers were compelled to accept time and motion studies and therefore had to take the same road as the workers in the metal trades: increased information, increased enlightenment. The study activities for local union officials of the Textile Workers Union started in 1935.

The Textile Employers Association also had to do something about the increasing complexity found in the interplay between conditions of production and the formulation of the text of the agreement. The setting up of the technical conference form 1936 is an expression of this.

At the beginning of this investigation the question was asked about when the texts of the agreement and the physical conditions of production became so complicated that a so-called engineer's expertise became unavoidable for dealing with them. Since, as the investigation has shown, both development of the agreement and conditions of production progressed by degrees it is impossible at this aggregational level to supply any exact answer. But it is clear that it was during the latter part of the 1930s that concrete measures were taken on both sides to increase the theoretical qualifications of those who were to handle the system for the setting of piece wages, which presumed time and motion studies. It was during the same period that these studies obtained their spread within the industrial branches concerned with increasing speed. There seems to be some causal connection here. This connection could somewhat elliptically be expressed thus, that the introduction of time and motion studies into the setting of piece wages not only gave the trade union organizations an increased influence at the cost of the employers, but also gave the trade union functionaries who could meet the demands for higher qualifications an increased responsibility for the workers who had been deprived by the development of even more of their independence.

But it was not only an increased need of a general technical competence that contributed to estrange the agreements on piece wages from the workers and transfer them to the sphere of the trade union representatives. Every extension of the area covered by the written national agreement also carried with it an extension in influence of the lawyers. The risks then threatened were quite clear to the workers' movement. The opposition to legislation about trade agreements and labour courts was based on these fears. At headquarters the leadership soon accepted the legislation and proceeded further on this basis along the line of compromise with the employers and representatives of the public. But this legislation increased the demands for precision and carefulness in the work of negotiation. The compulsion to submit the agreements made to the test of interpretation when disputes arose was taken by K. J. Olsson as justification for the necessity of terminating the textile agreement in 1930. It was necessary to see that the texts of the agreement were so constructed that the gains that were considered to have been harvested at the negotiation table also withstood the test of those skilled in the law. But it was a long time before the workers could free themselves from the suspicion that the court basically served the interests of established society and the employers.

Against this background it is interesting to observe the role of the Labour Court in the development of the balance between the parties in relation to piece wage questions and time and motion studies. In the limelight here are the judgments noted in this investigation about the metal trades agreements of 1932, 1934 and 1944 and the judgment in the Rydboholm case in 1934. Initially it was presumed that the Labour Court—given the theoretical relationship between the ideas forming the ground for the court and the collective agreement law on the one side and the ideas of the rationalization movement on the other—would encourage the kind of solution of conflicts that increased the utilization of intersubjective criteria, including time and motion studies. The four court actions related in this investigation are not at variance with this assumption.

In the dispute of 1932 it was necessary for the court to decide on interpretations put forward by the parties, derived from the anxiety of the employers concerned to safeguard pay mobility and the workers' demand to increase security of earnings. It has been observed that in this situation the court acted in an independent manner and selected a different path from those immediately suggested by the parties. In this independent solution the court gathered many considerations together in a farseeing manner. They admitted the employers' demands for increasing the efficiency of industry, the necessary mobility of the nominal piece wages rates but at the same time linked up with the line positive to rationalization in the Metal Workers Union and admitted the workers' right to undiminished earnings. The result of the judgement given by the court was that rationalization and the utilization of time and motion studies were stimulated. This was underlined not least by the judgment of 1934 when the court confirmed that security of earnings applied to the individual not to whole plants or companies.

Completely in line with this lay the attitude to the struggle of the Rydboholm workers against job studies as a basis for piece wages. Unanimously the court confirmed the impossibility of the workers to obstruct or refuse participation in time and motion studies.

The decision in the STAL case of 1944 meant that the line taken by the court from the beginning of the 1930s was carried further. Once again it was confirmed that the earnings-level for piece wages in the metal trades industry was to be safeguarded at a time of increasingly vaster rationalization. This was to occur through a guarantee of the individual piece wage but not for the individual worker, who thereby in the setting of new piece wages could not count on receiving a built up piece wage level or leave such as an inheritance to this successor. The argument of the Labour Court did not proceed from a person but meant a marked concentration on the not personally attached, "the objective system". A further emphasis on the way the court envisaged the further development was the clear indication of the general purpose piece wage formula in the metal industries. On this point the court's decision confirmed and stressed the development towards centralization and professionalization of piece wage setting as was described above.

Latter day opinions with a time perspective from the end of the 1920s up to the 1970s have evoked the scenario of the Labour Court as firmly established in an idea of justice that lay close to the employers and conservative to an increasing degree, bound by its own praxis and precedences.[4] Thus, for example, writes Sten Edlund in a contemporary characterization:

... seems unfortunate that the Labour Court was so hard and self-sufficient in the traditional retrospective court manner instead of ranging itself as a flexible element in a larger system in the course of development.[5]

Against the background of the actions of the Labour Court shown here this assessment does not seem completely adequate. At the beginning of the 1930s when of course the court had no internal tradition to support it the judgment instead, precisely because of its independence that perhaps sometimes deserved to be called self-sufficiency, appears as a progressive force in the area of high tension created by the relations between the parties in Swedish industry during a period of accelerating transformation. From the perspective established here it is the role of this court in the breakthrough stage of modern production technology and organization that is interesting and relevant. Then the role of the Labour Court becomes simply a part of a greater system undergoing development. The Labour Court's interpretation of agreements fitted well into the demand of modern production technology for the security of earnings that did not restrain the increased efficiency of industry. It can hardly be maintained in view of the line taken by the Trade Union Confederation and the trade unions from the middle of the 1930s, that in these matters the Labour Court devoted itself to some kind of blind support of the employers' interests. Instead the court joins with the agencies and organizations that in the development of scientifically permeated compromise solutions saw the possibilities of social development.

8. Conclusions and new points of departure

The dominating question in this study was how a changed production technology influenced the relationship between the parties of the labour market. The investigation of the complex problems of time and motion study in the agreements of the textile and metal trades areas in Sweden during the period 1920 to 1950 should be regarded as a case study relating to the general problem.

The techniques of time and motion studies and the rationalization measures of which these were a part, caused or strengthened tendencies to changes in the balance on at least three levels or along three dimensions.

The first of these is the balance of strength between workers and employers as parties. Rationalization carried with it shifts to the advantage of the former.

At the same time production technology became, as the conditions for the sharing of profits by wages, a complicated thing. Experts with different aims increasingly took over more of what earlier could be combined with the traditional roles in the working community. This can be regarded as professionalization, a shift along the dimension of layman-professional.

The third dimension links up with the sociological core—periphery—models. For practical reasons professionalization carried with its a shift in influence from the individual to the organization and within the organization from the local agents of the organizations at the periphery to the central heirarchy of the representative.

It seems necessary to consider the effects of technology on society with the help of such a multidimensional presentation. At the same time it demonstrates the difficulties of giving clear answers to such simple an reasonable questions about which party gained from the introduction of the new technology. If the shift mentioned first is adjudged as democratization, a step in the direction of economic democracy, this retards the empression of disturbances along the dimension of professionalization. It was not the workers who gained in influence but their organizations. The individual employer seems to have lost most in influence, to the workers and to his own organization. A summary of the results obtained from the investigation may be expressed in the following manner.

Modern production technology, whose characteristic features also included the technique of time and motion study, had the effect during the period 1920 to 1950 that the workers' side increased its influence (its power) in relation to the employers' influence and this growth occurred at a professionalized level close to the centre of the organization.

As a starting point for continued investigations the experiences gathered during this investigation suggest an approach to the problems surrounding the relationship between technology and bureaucracy. Is the growth of bureaucracy—in organizations, in government and the municipalities— a necessary result of rationalization (the turn to mathematics) and of production and administration? What alternatives have there been? How conscious have the actors been about such resultant effects and in such case were they prepared to accept them? Would the earlier advocates of the rationalization movement, men like Erik August Forsberg and Alex Engblom, have spoken so warmly for their case if they had been aware that precisely through this they had contributed some of the soil for the marked growth in power of the trade union movement? What assessments of the consequences have other decision makers had in the long list of successive decisions that have required increasingly sophisticated technology in order to solve social problems and to achieve social solutions? New questions crowd in at the conclusion of this presentation.

Appendices

Appendix 1. Certain texts from the piece wage paragraph in the metal trades agreements 1935–1948

1. Re job studies

(New features in relation to the preceding agreement are in italics)

1935 para. 4. Clause 4.

In the so-called premium piece wage, prior to its introduction, the grounds for this shall be confirmed and agreement on it reached in due form.

1936 para. 4. clause 4.

Similar wording

. . . in due form.

In cases where work performances are studied, the worker concerned has the right to be informed about the grounds for the system used there as well as the result of the job studies.

1938 para. 4. clause 4.

. . . in due form.

In the cases where the *work* is job studied, the *workers* have the right to be informed about the grounds for the system used there as well as the *job study*.

1940 para. 4. clause 4.

As above

1943 para. 4. clause 4.

As above

1944 Para. 4. clause 4.

. . . the workers have the right *and, if they so desire, the representative for the works club* [the local union as represented in a factory; authors note] to be informed about the grounds for the system there used as well as the job study.

The same applies with regard to diagrams and tables to the extent they form the basis for the piece wage agreement. If, in the event of a dispute, a control study is arranged, the representative of the works club has the right to follow this.

The right that in accordance with the above is ensured to the representative of the works club shall, where the dispute regarding the piece wage question gives rise to central negotiations, likewise apply to the representative of the main organisation.

Minutes:

1. At companies where the piece wage is fixed on the basis of job studies, the worker also has the right to propose that a certain job shall be job studied. The worker's desire thus advanced should, to that degree the worker can provide reasons for this, be complied with without unnecessary delay.

2. In the assessment of the scope of work studies regard should be paid to the desires put forward by the worker there concerned.

3. The parties are agreed on recommending the employer and the employee at the company concerned that, where such agreement does not already exist, to try to reach agreement on the grounds for calculation.

4. It is presumed that the parties concerned cooperate loyally to obtain a correct result from the job study work.

1945 para. 4. clause 4.

As above

Minutes:

1. . . . The worker's desire thus presented *should be complied with without unnecessary delay whilst noting that the worker with regard to the work concerned, which has been job studied earlier, must provide a reason for his request.*

. . .

3. *At companies where agreement on the grounds of calculation does not already exist, the employer and the worker should make such agreement.*

4. . . .

5. *In order to promote reliable cooperation and good working conditions and also for the adjustment of differences in everything concerning job studies and the setting of piece wages a job study committee shall be set up consisting of four ordinary members and five deputies. Of these, two ordinary members and two deputies are to be appointed by the Metal Trades Employers Association and two ordinary members and three deputies by the Metal Workers Union and the Union of Foundry Workers jointly. The committee shall appoint a chairman, a vice chairman and secretary.*

When the committee considers it necessary or when it is requested by either party, the committee undertakes the necessary investigations and thereafter reports to the parties concerned as soon as possible the results of these and gives the instructions required by them. It is presumed that where full agreement exists within the committee to give certain instructions, these shall be followed by the parties concerned. Measures in accordance with what has been said above should be carried out as quickly as possible.

Impending measures do not carry any restriction in the right of any party to demand the treatment of disputes in accordance with the procedures laid down in the agreement.

1947 para. 4. clause 4.

As above

Minutes:

. . .

3. . . . make such agreement. *Where either party so requests, local negotiations shall be instituted as soon as possible, and, if the local negotiations cannot be concluded within two months, the question may be referred to central negotiation.*

4. When a dispute arises on a piece wage that has been job studied, the main organizations on request may receive copies of the job study report. Such docu-

ments are to be regarded as an internal matter within the company and may not be supplied to those not concerned.

5. = minute 4 of the preceding agreement.

6. = minute 5 of the preceding agreement.

1948 para. 4. clause 4.

As above

2. Re the grounds for calculation of piece wages

1948 para. 4. clause 5

a) The calculation of piece wages *in connection with job studies* shall in principle be carried out according to the following formula
$x=t\times u\times(1+s/100)\times p/60$,

where x = the piece wage in öre, t = selected effective time in minutes, u = performance index, s = lost time bonus in per cent and p = money factor in öre/hour in accordance with the table below. The selected effective time is calculated according to the "average value method". The performance index is calculated with reference to the worker's performance during the job study (u = 1 for normal performance by an average skilled worker, greater than 1 if a performance is higher and less than 1 if it is lower.) The lost time bonus is calculated with reference to the necessary lost times for the work in question.

		Money factors in öre/hour for appropriate cost of living group				
	Worker group	1	2	3	4	5
A	Skilled workers in accordance with par. 3 clauses a & b	170	178	187	195	210
B	Other male workers aged 20 and above	155	160	169	175	185
C	Male workers aged 18/19 and female workers aged 20 and above	112	120	125	130	140
D	Male workers aged under 18 and female workers under 20	90	95	100	105	110

b) *When the work is not job studied,* the above money factors are used in the setting of piece wages as targets in the normal performance of an average, skilled worker.

Minutes:

1. By the introduction of these rules is not intended any alteration in the provisions in clauses 2 and 3 of this paragraph.
2. The above grounds for the calculation of piece wages may not be taken as a pretext for demands about alteration in the previously existing piece wages.
3. Grounds for the calculation of piece wages agreed earlier shall cease to apply in the setting of piece wages that takes place after the 6 weeks immediately following the approval of this agreement.
4. The provisions in this clause do not create any hindrance for the main organi-

zations in exceptional cases, where special reasons exist, to permit the use of other grounds for the calculation of piece wages.

Appendix 2. Certain texts from the piece wage paragraph 7 of the textile agreement 1914–1951

1914

1. At every factory for the work that is carried out on piece wages price schedules shall be set up, which apply for mutual observance and for the same period as this agreement.

2. If new working methods or improved machines are introduced for the work or if changes in the work or the article are introduced, adjustment in the previous piece wage rates may occur; however, such new piece wage price may not be set lower than that which a worker with the same skill and industry can thereby achieve the same average earnings weekly as before the change.

3. Piece wage work may otherwise be carried out in all of those cases where the employer considers this suitable when, for all piece wage work that has not been decided by means of fixed piece wage schedules, agreement on the price will be made by free negotiations in every case between the employer or his representative and the worker or workers to whom the piece wage work is offered. If no agreement on the piece wage price can be made, the work shall be carried out at hourly pay.

1920

1. (the same as the preceding agreement)

2. If new working methods or improved machines or if changes in the work or product are introduced, the piece rates affected by this may cease to apply one week after notice has been given. When a new piece rate is desired agreement about this shall be made in accordance with the provisions stated in clause 3 here below.

3. ... in each case between the management or its representative ... to whom the piece wage work is offered. Should an agreement not be reached between them the matter may be referred to the organizations. If agreement is not reached even then the work shall be carried out at hourly pay.

1922

1. At every factory, for the work that is to be carried out on piece wages, piece wage schedules shall be drawn up, which shall be mutually observed and for the same period as this agreement.

The piece wage prices shall be so set that they provide a worker with an excess above the hourly earning of an average 15%. The piece wage for such work in which both male and female workers are engaged is calculated on the hourly pay and production of the females when the male workers shall be given the opportunity of making up their earnings by being given work at larger, heavier or more machine units or, where this is not possible, in some other manner preferably by premium piece wages or a certain bonus per hour. In so far as the piece wage does not refer to underaged workers the piece wage price shall be calculated on the hourly pay for the highest age class stated in the pay tables.

2. (As the preceding agreement.)

3. (As the preceding agreement.)

1925

1. *Fixed piece wages laid down in schedules*

a) At every factory piece wage schedules shall be drawn up for the work carried out at piece wages, which apply for mutual observance and for the same period as this agreement.

Piece wage schedules shall be drawn up in the manner laid down in the "negotiation procedure for the drawing up of piece wage schedules", agreed between the organizations.

b) If new working methods or improved machines or changes in the work or the article are introduced, the piece rate affected may cease to apply one week after notice has been given.

Where another piece rate is desired instead of that for which notice of termination has been given, agreement about this shall be made in accordance with clause 2 in the provisions given here below.

c) If during the currency of this agreement a piece wage is added to the piece wage schedule, this question shall be submitted to the organizations for approval.

2. *Other piece wages*

Piece wage work may otherwise be carried out in all of those cases where the management consider it suitable. Agreement about the price ... (the continuation is the same as in the applicable parts of the preceding agreement, clause 3).

3. *Grounds for calculation of piece wages*

Piece rate ... age class. (As the previous agreement, clause 1, passages 2 and 3.)

1931

1. *Piece wage schedules*

a) At every factory ... observed where agreement on a shorter period of validity has not been made, for the same period as this agreement. The drawing up of a piece wage schedule shall be carried out in the manner laid out in paragraph 17 here below with reference to adjustment of disputes. The piece wage schedule shall be provided with a note to the effect that it belong to this agreement and be signed by the employer and the trade union concerned and be approved by the main organizations.

b) If new methods or improved machines are introduced or changes in the work or the article are introduced, notice of termination of the piece wage to cease in one week may be given by either party. A report on such notice shall be made by the parties to the main organizations without delay. Where another piece rate is desired ...

c) During the period of the agreement in accordance with b) above the agreed piece wage shall apply for the same period as the national agreement, unless agreement has not been reached on a shorter period of validity. A piece wage that has been fixed for the same period as the national agreement shall be submitted to the main organizations for approval and then added to the piece rate schedule.

2. *Other piece wages*

Piece wage work may ... offered. If an agreement between them cannot be reached the work shall be carried out at hourly pay, the parties however being at liberty to appeal in accordance with the rules stated in paragraph 17.

3. *Grounds for calculation of piece wages*

a) In the establishment of a piece wage in accordance with clauses 1 and 2 the piece wage shall be set so that under normal working conditions it shall give an average worker an excess above the hourly pay of at least 15%, when special workers, who are given more responsible work, either heavier such or with regard to more machines, or which requires specialy accuracy and skill, should receive a higher excess.

b) Piece rate ... age class. (As the preceeding agreement, clause 1, passages 2 an 3, but not the first sentence in passage 2.)

c) If an investigation is requested from either party concerning the production that shall form the basis for the setting of a piece wage, such investigation shall be carried out. The investigation shall refer to average production for an average worker. The management and workers shall in consultation select the workers on the basis of whose production the piece rate shall be set, and the workers' side shall be given the opportunity of taking part in or checking the investigation.

The investigation may not proceed for a longer period than is required to obtain a satisfactory investigation and may not in any case continue longer than one month.

1937

1. a) At every factory a piece wage schedule shall be drawn up for the work carried out at piece wages. Provided an agreement on a shorter period of validity has not been reached, this schedule shal! apply for the same period as this agreement. A schedule with the last named period of validity is called the fixed piece wage schedule and the piece wages there listed are called fixed piece wages. Piece wage schedules with a shorter period of validity are called provisional piece wage schedules and the piece wages therein provisional. The establishment of piece wage schedules, either permanent or temporary, shall be carried out in the manner and in the way laid down in c) here below. The schedule shall be signed by the employer and the appropriate trade union. Furthermore a fixed piece wage schedule shall additionally be approved by the main organizations and provided with a note that it belongs to this agreement. If central negotiations have not been requested within a month after one of these organizations has notified the other of the agreement then such schedule is considered to be approved. The piece wage schedule may not contain provisions contrary to this agreement.

b) Introduced ... in accordance with the provisions given here below under c).

c) Piece wage work may be ... offered. But where the question is about piece wages that shall be introduced into the piece wage schedule, the worker shall be entitled to turn to his trade union in the matter. If agreement cannot be reached between them, the work shall be carried out at the piece wage offered by the employer or at hourly pay, the parties however being at liberty to appeal in accordance with the rules laid down in paragraph 17 concerning the adjustment of disputes.

Agreement on the fixed or temporary piece wages may be applied retroactively if agreement is reached within three months after the piece wage is issued. If the management, in cases mentioned in b) here above or where the question otherwise is about piece wages in such groups of work that are usually regularly carried out at piece wages, wishes to postpone the setting of a piece wage, the worker is entitled until the new piece wage is offered to receive in addition to the hourly pay a bonus of 10%. Piece wages, which are not drawn up in writing or which are established without the assistance of the trade union only apply until further notice and subject to one week's notice from either side.

2. a) (As the preceding agreement.)

b) If the piece wage price for such work in which both male and female workers are employed is calculated on female hourly pay and production, the male workers shall be given the opportunity to increase their earnings by being given work at larger, heavier or more machine units or, where this is not possible, in some other manner, preferably by premium piece wage or a certain hourly bonus.

So far as ... (as the preceding agreement).

c) (As the preceding agreement.)

1939

1. a) (As the preceding agreement)

b) (As the preceding agreement)

To the clause is attached the following minute:

The parties are agreed that on transfer to a greater number of machines, notice should be made of the old piece rate in so far as it refers to the new working conditions.

In a special regulation in connection with prolongation it was laid down:

In the application of paragraph 7 clause 1b of the agreement for textile workers the parties are likewise entitled to cite steps taken prior to the coming into force of that agreement. (This provision becomes a minute in the following agreement.)

c) passages 1, 2 and 4 are as the preceding agreement. Passage 3: Desire ... a bonus of 15%. Clause 2 as the preceding agreement, but with the difference that to the first passage of 2c has been added a reference to minute 3 of paragraph 12 clause 2: With regard to such stoppages in the work caused by the falling of a beam, the parties are in agreement that the four hours—calculated from the start of the stoppage—for which waiting time bonus shall not be paid, shall be included in the working time in the calculation of weaving piece wages where these occur.

1947

Clauses 1a and b as the preceding agreement. Clause 1c as the preceding agreement with the following alteration in passage 3: Desires ... draw pay for the amount stated in point 2 a) stated below.

Clause 2a. In the establishment of piece wages in accordance with clause 1) piece wages shall be so set that under normal working conditions thay shall provide an average worker with at least the stated hourly earnings in öre given below, when special workers who are given more responsible or heavier work or with regard to more machine units or that require special accuracy or skill, should have an hourly wage that lies higher than what is thus stated.

Schedule	1	2	3	4	5
Male workers	161	155	152	146	141
Female workers	125	123	121	117	115

Clauses 2b and c as the preceding agreement.

1951

Paragraph unchanged with the exception of the following passage in clause 1c:

...

If agreement cannot be achieved between them, the work shall be carried out at the piece wage offered by the employer, either party still being at liberty to appeal in accordance with the procedure laid down in paragraph 17 of the main agreement mentioned. When the question of the setting of piece wages has been referred to negotiations between the main organizations then, where it has not been possible at the main negotiations to achieve any agreement in the piece wage question, payment shall be made at the hourly wage, unless otherwise agreed.

...

Appendix 3. The Labour Market Committee's proposal about job study organs. General motives

The general and important condition for the improvement of the living standards of the workers and the whole people is a continuous increase in efficiency of trade and industry. This increase in efficiency must be continuous in every company so that the company shall be able to keep pace with technical development. It is of common interest for all of those employed at a company to cooperate so that it is able to meet competition. As every other member of society the employees profit by a rationalization, which leads to increased production and thus to that degree creates a basis for a general increase in the standard of living. As an important factor in this rationalization are included measures for the purpose of relieving production from unnecessary costs for material and labour. Correctly carried out, job studies are a rational tool for the achievement of this purpose.

In Sweden job studies started to come into more general use at the middle of the 1920s. On the part of the workers they have sometimes been met by opposition or suspicion, to a higher degreee earlier than at present. There have been many resons for this. Mistakes and wrong judgements have obviously been made from both sides. An increased measure of collaboration between the parties of the labour market with regard to job studies seems to a great extent to be aimed at ironing out the differences that still exist in this area.

It is important that the possibilities of local collaboration in questions of job studies are utilized. Preferably the local collaboration should occur in the shape of discussions between the employer and the local trade union organization of the workers at the company (works club or such); disputes about job studies that are of immediate importance for the setting of wages should always be dealt with in this manner. Questions of principle concerning the arrangement and carrying out of job studies may also be dealt with in the works councils, set up in accordance with the agreement of 30 August 1946 between the Swedish Employers Confederation and the Swedish Trade Union Confederation. At companies where a company union representative has been appointed such questions could instead be dealt with in collaboration between employer and the union's company representative.

In order to increase collaboration in job study questions still further the central organizations have made the agreements which follow on the setting up of a work study council and of work study committees within the spheres of activities of the associations.

Job studies (time and motion studies) are tools to discover the simplest and most efficient method to carry out the work and to obtain reliable grounds for the setting of piece wages. Job studies consist essentially of the gathering and processing of information about the work, investigation of the work place, technical facilities, and so on.

Such studies are thus a matter of production engineering. But the studies may be misleading and produce faulty results if attention is not also given to physiological and psychological factors. So during a study the worker should use his normal working speed. The collaboration of the worker himself in the arrangement and carrying out of such studies therefore constitutes an important condition for a satisfactory result of job study activities.

During the studies the job study man should take note of the views put forward by the workers. Obviously he shall also collaborate with the supervisors and foremen who are immediately responsible for the work. Orders and instructions of great importance for the workers should be made through the supervisor or foreman concerned and not through the study man. In his investigations the latter should always observe occupational hazards and try to satisfy views about safety devices, lighting, ventilation, etc. and even, so far as is possible, try to make the work easier for the worker. It is quite natural that such studies do not always lead to immediately obvious results. Therefore it is important that the workers are informed about the purpose of the studies. In this way the workers obtain a better chance of putting forward proposals for improvements themselves.

In the calculation of piece wages, job studies are utilized as a tool for the purpose of setting the piece wage so justly as is possible. In this it should be seen that the adjustment of earnings to the individual worker's diligence and skill is not prevented.

Before a piece wage based on job studies is offered to the worker the job study man should have made sure that the work is carried out so that the earnings' potential of the worker is, so far as possible, not influenced by shortcomings in the organization of the operation or the technical equipment. In order to avoid faulty generalizations it is important that so-called lost time is weighted with reference to the kind of work and is in so far as possible based on job studies.

The setting of piece wages shall be carried out in accordance with the provisions laid down in the current collective agreement. In negotiations because of a dispute about piece wages, as a rule the job study man should only participate as an expert in order to present the requisite information and investigations for the parties concerned. Where at present in accordance with praxis the study man takes part in negotiations also as the representative of the employer or where this procedure is caused by conditions among the personnel at the company, there shall be no obstacles to the use of such procedure.

High demands must be made on the competence of the job study man. He should have a technical training or comparable knowledge gained by practical work. He should be practically inclined and have a good experience of work within the branch. He should have reached such a degree of maturity and have such personal characteristics that in an objective, impartial manner to inspire confidence, he can fulfil his task. He should carry on his studies openly and when so required inform the worker about the study material. It is of importance that job study men in the performance of their task receive the support of the parties at the workplace.

Through the following agreement the Swedish Employers Confederation and the Swedish Trade Union Confederation undertake to carry out their duties in the work study council as rests on them according to the agreement and to act so that the agreement becomes accepted as a collective agreement between the associations belonging to them by free agreements. To the extent that this occurs the agreement becomes otherwise also binding for the Swedish Employers Confederation and the Swedish Trade Union Confederation with legal effect in accordance with the law on collective agreements.

Concerning work study committees

Paragraph 1

Associations, who have between themselves accepted this agreement, shall set up a common committee on questions concerning the arrangement and carrying out of work studies (work study committee) with the task

of encouraging local collaboration between employer and worker on job studies

and also in the procedure set out below to deal with disputes about job studies.

Paragraph 2

The work study committee consists of four members of which two are appointed by each party. An equal number of deputies are appointed for these members.

The committee appoints a chairman and vice chairman and a secretary, within the committee or outside it.

The committee shall decide about meetings and procedure.

Paragraph 3

It is the task of the work study committee, on the application of an association that is a party to the agreement to deal with disputes concerning the arrangement and carrying out of work studies. In so doing the committee shall, if so requested by either association, undertake the necessary investigations at the work place involved in the dispute.

The provisions in this paragraph do not carry with them any limitation in either party's right to call for central negotiations for the solution of the dispute in the otherwise current procedure for this. If such negotiations have been requested the dispute may not contrary to the objection of either party be taken up by the committee before the negotiations have been concluded.

Paragraph 4

The disputes envisaged in paragraph 3 shall be dealt with by the committee with all possible speed.

The result of the committee's examination with the instructions following therefrom shall be delivered in writing as soon as possible to the parties concerned. When within the committee unanimity has been obtained for a certain instruction, the associations shall work for this to be complied with.

Concerning the work study council

Paragraph 5

The Swedish Employers Confederation and the Swedish Trade Union Confederation appoint in conjunction with the acceptance of this agreement the joint organ, the work study council, with the task

of following and encouraging collaboration with regard to work studies and work study committees' activities and otherwise work for a sound and useful organization of work study activities

and to as a conciliation board take up and decide disputes about the validity or the correct meaning of the provisions in this agreement, and also whether certain procedures are contrary to these provisions, and also about the penalties for procedures found to be contrary to them.

The council consists of three members from each of the organizations and the same number of deputies for them. The members and the deputies are appointed for a period of three years.

The Swedish Employers Confederation and the Swedish Trade Union Confederation appoint for a period of three years a neutral chairman to serve at the council's handling of cases, which it has to decide as a conciliation board.

Paragraph 6

There will be a secretary at the council for both the Swedish Employers Confederation and the Swedish Trade Union Confederation.

This agreement shall between the Swedish Employers Confederation and the Swedish Trade Union Confederation as well as between the associations, which have accepted it, apply until further notice with a term of notice of six months.

Saltsjöbaden 26 May 1948

Sven Schwartz
Elof Ericsson
Axel Enström
Arvid H. Nilsson
Erik Bengtson
Bertil Kugelberg
/Alvar Odholm

Axel Strand
Gust Wahlberg
Oskar Westerlund
Ivan Larsson
Hj Adiels
Einar Norrman
Ernst Falk
/Arnold Sölvén

Notes

Chapter 1. Problems and sources

1. De Geer 1978.
2. Segerstedt-Lundquist 1952, p. 167 et seq.
3. In their industrial-sociological pioneer work "Människan i industrisamhället—Arbetslivet" (Man in industrial society—Working life), *Torgny Segerstedt* and *Agne Lundquist* have investigated a number of factors that have influenced mankind's adaptation to the "new" industrial system of work. The field work in the shape of questionnaire investigations was carried out at the end of the 1940s in a number of industrial companies and industrial districts. Job studies belonged to the phenomena where the attitudes of the workers were investigated.

Segerstedt proceeded from a partly informed attitude to rationalization and job studies found in spokesmen for the trade union movement. He there found that with the passage of time the so-called general pattern in this respect had undergone a change from a negative to a positive attitude. Against this background he expected to discover differing attitudes, for example between older and younger workers on the factory floor. The researchres discovered that there was a rather good association with the general pattern of the workers' attitude in the job study question. The union leadership thus succeeded in making their ideas clear to the members. There was one factor, however, that could decisively break this relationship and which therefore appears as, apart from the view of the trade union management, the most important control variable. Those who had themselves been subjected to job studies tended to be markedly more negative to job studies than those who had not personally been subjected to them.

There was a possiblity of differentiation even within the studied groups. To have an average skilled job at medium high level of mechanisation and be subjected to job studies was the combination of conditions that principally created the most negative attitude to job studies (p. 177).

In the volume "Information på arbetsplatsen" (Information at the workplace), 1956, *Edmund Dahlström* discussed the structural communication problems within companies. The purpose of his study was to present research problems and discuss them and was arranged as a survey of the research position, followed by three intensive studies and a concluding discussion. One point of departure for the investigation was found in the intersection between a structural line of development of industry in the direction of increasingly complex systems, and a political trade union one directed towards a democratization of working life. One of the problems distinguished in the tense situation that had arisen was the setting of piece wages and job studies. The communication between the actors in this respect was made the subject of study. Two companies, a smaller and a larger, in both cases with good relations between management and employees, were selected for case studies. A field interview investigation was carried out mainly during 1954.

Dahlström's book provides a good picture of the moods and attitudes surrounding job studies in the people affected at a workplace: the workers, the study men, the study managers, and the management. In these attitudes are shown a number of possible explanations of the reactions that found expression not least in union demands and aims. It should be noted, however, that Dahlström's investigation reflects conditions at the middle of the 1950s, thus following the period covered by the present investigation. During the intervening years, the first half of the 1950s, also occurred the introduction of a more refined synthetic system, including time and motion studies, which caused increased irritation and a more comprehensive public debate. It should also be noted that Dahlström does not investigate specially the framework of the trade union organizations and their actions in connection with the setting of piece wages.

A document from the period at the middle of the 1960s was the report to the Swedish Trade Union Confederation's congress of 1966, "Fackföreningsrörelsen och den tekniska utvecklingen" (The trade union movement and technical development). Behind this stood a working party in the general council of the Trade Union Confederation under the chairmanship of Rudolf Meidner. Among the experts attached to the group was again found Edmund Dahlström. The working party sided with the traditional evaluation in the Trade Union Confederation about the need for a continuous rationalization of working life, and this idea created a framework surrounding the deliberations of the group (p. 23). Even the contribution from the trade union movement was assessed positively: "When the purely material consequences of the technical transformation and security of employment are concerned, the trade union movement up to now has largely been able to a reasonable degree to assert the interests of the members" (p. 216). Thus no reassessment of this foundation was called for. The report resulted in a proposal for a trade union action program dealing with the influence of technical development. The importance of further expertise and evaluations in respect of job studies was stressed. In addition the working group indicated the possibilities of utilizing the technique not only for the classification of working processes but also for the combination of work operations in a new manner: job design (p. 230 et seq).

With technical development was understood a comprehensive concept, almost a process of change in trade and industry as a whole. In the following certain observations and items in their report will be dealt with. A summarized description of the situation of job studies in Swedish trade and industry around 1965 provides some perspective in the extension of the present investigation (p. 157 et seq). The working group tries to capture the technical development in trade and industry in a structure of three levels of mechanization, originally developed by Robert Blauner. It was at the intermediate level that the importance of piece wages and job studies increased considerably. What special feature does this show?

At the intermediate level of the mechanization scale the contribution of the individual worker lost importance in comparison with the handicraft level. Machines or combinations of machines carried out an increasing proportion of the physical work and in addition took over through certain programming or regularity part of the decision function of the worker. To an increasing degree the role of the worker becomes that of the operator who started the machining, looked after it and took care of the product. The various sequences of operations were still short and repetitive. Longer processes and automation had their place in a high level of mechanization (p. 35 et seq).

The relationship between the level of technical development and kinds of pay, which is the focus of the present investigation, was nevertheless far from unambiguous or clear. The actions of the parties of the labour market were of great importance in this context, even though these factors were also dependent on the conditions created by the level of development of physical production (p. 172 et seq.) Here the working party noted a trend in development comparable with the results in the present investigation:

"From having had a position earlier where the employer wanted to introduce job studies and piece wages but the union side was doubtful, we now have a tendency to the opposite: the employers want to have greater control over and decrease wage dirft whilst the unions not so seldom see the chances of special increases in pay in a relatively free piece wage system.

The professionalization aspect was also observed in the report, both with regard to the demand to shoulder the expert role against the employers in negotiations as well as in the relations between the workers and the union representatives (p. 205 et seq).

The report's assessment with regard to cooperation and mutual understanding is interesting—that the increased tempo of changes tended to intensify the relations between the interested parties in a company. The conditions should thus not deteriorate because of changeovers. It is worth observing whether the same relation also applies at the level of the main organizations.

Whilst these words are being written the research situation in sociology is difficult to survey for an outsider. Overlapping themes such as technique and social change seem of latter years to have tempted an increasing number of researchers, and different aspects of themes such as questions about the working environment and job satisfaction (cf Gardell 1976) or problems concerning operational processes and job classification (see for example Social forskning (Social Research), Nos. 2–3, 1977) have been developed.

It is my impression that the trend in the latter case has increasingly moved in the direction of theoretical calculations and contributions to an international discussion whilst the empirical basis has become comparatively weaker or reflected development in other industrial countries than Sweden, especially the USA and Germany. Studies along the historical dimension in Swedish social development are conspicuous by their absence or are possibly being written. So far as time and motion studies are concerned I have not discovered any special studies.

4. In a similar manner Ulf Olsson, for the parties of the Metal Trades industry, has proceeded from one point in two constructed aims, summarized in the following way: "The Swedish Metal Trades Employers Association has mainly worked for two aims in wage policy: a large wage difference between different groups of workers and great freedom for local and easily changed wage setting. The Swedish Metal Workers Union has upheld two main principles in its wage policy: the equalization of differences in pay for different categories of workers and the greatest possible fixing of wages at the central negotiations." (Olsson 1979, p. 49).

5. Gardell 1976.

6. In other presentations, for example in the report to the Trade Union Confederation's Congress of 1966, "Fackföreningsrörelsen och den tekniska utvecklingen" (The trade union movement and technical development), job studies are regarded more as a kind of procedure for change in the organization and therefore not dealt with in the report. In my opinion such a

division only produces unnecessary and confusing difficulties in drawing up limits.
7. Korpi 1978, p. 135 et seq.
8. Gardell 1978, p. 157 et seq.
9. Geijer-Schmidt 1958.
10. De Geer 1978, p. 183 et seq.

Chapter 2. The metal trades industry—a presentation

1. The primary material is kept in the investigation's archives, RA (National Archives) committee 593
2. The presentation mainly based on Styrman 1946
3. For the sake of clarity the Swedish designations are: general council corresponds to *överstyrelse,* executive board corresponds to *arbetsutskott*.
4. The presentation is mainly based on the history of the Swedish Metal Workers Union, published 1938–1977 in Swedish, See Lindgren 1938, Lindgren et al. 1948, Back 1963, Back 1977.
5. The Swedish designations were: for the larger assembly (translated throughout here as general council) *förbundsstyrelse* up to Congress 1936, and following this *överstyrelse*. For the smaller assembly (translated throughout here as executive board) *verställande utskott* up to 1936 and thereafter *förbundsstyelse*.
6. The presentation is based on and agrees partly with De Geer 1978, p. 340 et seq.
7. De Geer 1976.
8. Kennerström 1974, p. 12.
9. Kennerström 1974, p. 17 et seq.
10. Cf. motion No. 3 to Congress of Metal Workers Union 1929, minutes p. 58 et seq.
11. Ibid p. 9 et seq.
12. Congress of Trade Union Confederation 1931, minutes p. 212 et seq.
13. Ibid p. 214.
14. Back 1963.
15. Congress of Metal Workers Union 1929, minutes p. 59 et seq.
16. Ibid p. 66.
17. Ibid.
18. Ibid p. 70.
19. Ibid p. 69.
20. Ibid p. 73.
21. General council of Metal Workers Union, minutes 1/11 1930.
22. Ibid.
23. Back 1963, p. 187.
24. Back 1963, p. 220 et seq.
25. Back 1963, p. 187 et seq.
26. Congress of Metal Workers Union 1932, minutes p. 184 et seq., motion 138.
27. Ibid p. 185.
28. Ibid p. 186.
29. Analysis based on the printed national agreement published by the Metal Trades Employers Association.
30. Styrman 1946, p. 309 et seq.

Chapter 3. The Swedish Metal Workers Union and the pay guarantee: the Chairman's line

1. A detailed presentation of this in Back 1963, p. 117 et seq.
2. For a discussion of the conference's competence see Back 1963, p. 126 et seq.
3. Back 1963, p. 126.
4. Back 1963, p. 126. Minutes of policy conferences on the agreements 24 Sept. 1926. . . . Stockholm 1926.
5. Ibid p. 136.
6. Policy conference of Metal Workers Union 1927 on the agreement, p. 16.
7. Ibid p. 17.
8. Conflicts about rationalization measures had of course appeared earlier. The most spectacular conflict occurred in Olofström 1915. See Lindgren et al. 1948, p. 584.
9. Ibid p. 18.
10. Ibid pp. 22 and 30.
11. Back 1963 p. 138 et seq.
12. Policy conference on the agreement, Metal Workers Union 1928. See contributions by Ernst Ohlsson, Anton Sjö, Fritjof Ekman, Gunnar Andersson.
13. Ibid, p. 21.
14. Ibid p. 20 (A R Svensson of Husquarna AB).
15. Ibid p. VI et seq.
16. Back 1963, pp. 142 and 144.
17. De Geer 1976. Back has not noticed the connection but permits the proposal on the paragraph of the agreement about joint determination to stand completely unexplained. Back 1963, p. 142.
18. The course of the negotiations is analyzed in Back 1963, p. 145 et seq, and in Styrman 1946, p. 226 et seq.
19. In the recommendation to use piece wage work in the piece wage paragraph, the verb "may" is exchanged for the somewhat more compulsive "should" in order to meet the workers' demand for "shall". A piece wage bonus was introduced; however, not eh demand of the Metal Workers Union for 50% above the hourly pay but 10 öre above the minimum pay. Formal changes were also made in the clause of the paragraph dealing with piece wage schedules and also about piece wage surplus.
20. Back 1963, p. 278 et seq.
21. Back 1963, p. 48.
22. Policy conference on the agreement, Metal Workers Union 1930. Speeches by Axel Lundqvist, A. R. Svensson, Hugo Carlström, Ernst Ohlsson.
23. Ibid p. 31.
24. Ibid p. 32.
25. Ibid p. 32.
26. Back 1963, p. 280.
27. Styrman 1946, p. 234.
28. Back 1963, p. 280.
29. General council of Metal Trades Employers Association, 11/4 1931.
30. Technical drafting committee to the general council of the Metal Trades Employers Association, 25/4 1931. Appendix to minutes of the same council 27/4

1931. The committee consisted of the members of the Association's apprentices committee: E. A. Forsberg, G. Ahlroth, P. G. Graham, O. Hellman, E. Lindström, G. Styrman, S. Forster and G. Hellberg. A representative for Munktells Mekaniska Verkstad belonged for special reasons and Graham never took part in the discussions. The drafting committee met 22–23/4 1931. The discussions are documented in detail in the minutes and notes. The discussions dealt mainly with the piece wage question.
31. General council of Metal Trades Employers Association. 27/4 1931.
32. General council of Swedish Employers Confederation. 27/4 1931, draft, appendix to minutes of general council of Metal Trades Employers Association on same day.
33. Regarding direct contact in April see Oscar Westerlund's speech at the policy conference on the agreement in December; general council of Metal Trades Employers Association. 17/8 1931.
34. Ibid.
35. General council of Metal Trades Employers Assn. 17/8 1931.
36. General council of Metal Trades Employers Assn. 17/8 1931.
37. General council of Metal Trades Employers Assn. 4/9 1931.
38. General council of Metal Trades Employers Assn. 4/9 1931.
39. General council of Metal Trades Employers Association 29/9 1931. The minute actually states that the board decided to keep to the decision to give notice. When this was first decided in consultation with the Swedish Employers Confederation is not known.
40. Styrman 1946, p. 235; Back 1963, p. 284.
41. Back 1963, p. 280 et seq.
42. Policy conference on the agreement, Metal Workers Union 1931. See for example contributions by A. R. Svensson, Alfr. Pettersson, Gunnar Johansson, Bror Nilsson, C. V. Johansson.
43. Ibid p. 20.
44. Ibid p. 20, C. V. Johansson's statement; Back 1963, p. 291.
45. Back 1963, p. 295.
46. Back 1963, p. 295.
47. Back 1963, p. 307.
48. Lindgren et al. 1948, p. 756 et seq; Styrman 1946, p. 202 et seq.
49. Back 1963, p. 308 et seq; Victorin 1973, p. 144 et seq; Styrman 1946, p. 209 et seq.
50. Cf. discussion at general council of Metal Trades Employers Association, 27/4 1931.
51. Labour Court 1932/105, preamble.
52. Victorin 1973, p. 153.
53. Victorin 1973, p. 144 et seq.
54. See e.g. Geijer & Schmidt 1958, p. 346.
55. General council of Metal Employers Association 17/8 1931.
56. General council of Metal Employers Association 17/8 1931.
57. Policy conference on the agreement of Metal Workers Union 1931, p. 9.
58. Labour Court 1934/181.
59. Back 1963, p. 210.

Chapter 4. The Build up

1. Back 1963, p. 312.
2. Policy conference on the agreement, Metal Workers Union 1932.
3. General council of Metal Employers Assn. 8/9 1932.
4. Back 1963, p. 317.
5. General council of Metal Employers Assn. 7/9 1933.
6. General council of Metal Employers Assn. 21/9 1934.
7. Policy conference on the agreement, Metal Workers Union, 1934, p. 9.
8. Back 1963, p. 319.
9. Policy conference on the agreement, Metal Workers Union 1934. Contributions by Ernst Ohlsson, Anton Norling, Hartvig Johansson, Valter Nordström, Lennart Eckerström, K. A. Åberg, etc.
10. Policy conference on the agreement, Metal Workers Union, 1934, p. 19 et seq.
11. Policy conference on the agreement, Metal Workers Union, 1934, p. 29.
12. Policy conference on the agreement, Metal Workers Union, 1934, p. 30.
13. Back 1963, p. 321.
14. Metal Trades Employers Association's draft proposal 1934, archives of Metal Workers Union, series of negotiations with Metal Trades Employers Assn. (Negotiations, Metal Trades Employers Assn.).
15. Negotiation team's report. Archives of Metal Workers Union. (Negotiations, Metal Trades Employers Assn.)
16. Back 1963, p. 322. Compare the assessment on Gunnar Andersson's conciliant unyieldingness on p. 355.
17. Back 1963, p. 322 et seq.
18. General council of Metal Employers Assn. 19/3 1935.
19. Regarding date of appointment see letter Styrman-Edström 2/5 1936, archives of Metal Trades Employers Assn. Regarding Ingelman's biography: biography of the Swedish Assn. of Engineers and Architects 1936.
20. Congress of Metal Workers Union, 1935, p. 202.
21. Congress of Metal Workers Union, 1935, p. 206 et seq.
22. Congress of Metal Workers Union, 1935, p. 128.
23. Back 1963, p. 443.
24. Draft proposal: archives of Metal Workers Union. See Negotiations, Metal Trades Employers Assn.
25. Back 1963, p. 441.
26. Back 1963, p. 445.
27. Report about negotiations of the Metal Trades Industry, dated 7 January 1936, archives of Metal

Workers Union (Negotiations, Metal Trades Employers Assn.).
28. Back 1963, p. 447 et seq.
29. Report about settlement etc., dated 21 January 1936, archives of Metal Workers Union (Negotiations Metal Trades Employers Assn.).
30. Styrman first brings up the provisions about the right of the workers to information in connection with the 1938 agreement and is thus misleading on this point; p. 235.
31. Back 1963, p. 452 et seq.
32. Back 1963, p. 455.
33. De Geer 1978, p. 147 et seq. A summary of the discussions at the Royal Academy of Engineering Sciences (IVA) 26/2 1936 is attached to the minutes of the technical drafting-appretices committee, archives of Metal Trades Employers Assn.
34. Minutes of technical drafting-apprentices committee 27/2 1936.
35. Undated, unsigned memorandum attached to minutes of the technical drafting-apprentices committee of the Metal Trades Employers Assn. 27/2 1936. Presumably written by Gösta Ingelman.
36. Back 1963, p. 455.
37. Cf Meidner 1973, p. 8.
38. Cf Back 1963, p. 470, where the role played by Andersson is described as remarkable, as some kind of additional chairman.
39. Back 1963, p. 456; Olsson 1958, pp. 210 and 213.
40. Cf Back 1963, p. 456.
41. Policy conference on the agreement, Metal Workers Union 1937, p. 10 et seq.
42. Contributions by Edvin Nyström, Bertil Sundberg, P. A. Lindberg, Alf Johansson, Carl Rundberg, Georg Andersson, Sven Johansson, K. A. Åberg, Joh. Svensson, Sigfrid Wikström, Sam Jansson, Albin Öberg, Bror Hallengren, Gösta Johansson, Ernst Ohlsson and John Kristiansson.
43. Contributions by Bertil Sundberg, Carl Rundberg and Georg Andersson.
44. Policy conference on the agreement, Metal Workers Union, 1937, p. 21.
45. Policy conference on the agreement, Metal Workers Union, 1937, p. 21.
46. Back 1963, p. 364; Olsson 1958, p. 220 et seq.
47. Back 1963, p. 365 et seq.
48. Summary of the committee's report and conciliation proposal in archives of Metal Workers Union; Negotiations Metal Trades Employers Assn.
49. Back 1963, p. 246; Olsson 1958, p. 225.
50. Back 1963, p. 462.
51. Congress of Metal Workers Union 1938, p. 171 et seq, motions 83–90, which came from Svedala, Gothenburg (2 motions), Svalöv, Finspång, Huskvarna, Skövde, Stockholm.
52. Congress of Metal Workers Union 1938, p. 175.
53. Congress of Metal Workers Union 1938, p. 176.
54. Congress of Metal Workers Union 1938, p. 177.
55. Ragnar Nilsson, Svalöv, p. 180.
56. Gustav Vahlberg, Oscar Westerlund, Bror Nilsson, Sven Bergman, Marcus Stenhård, Evert Sträng, etc.
57. See also in Karl Berg.
58. Anton Sjö, Albert Johansson.
59. Congress of Metal Workers Union 1938, p. 188 et seq.
60. General Council of Metal Workers Union, 29/9 1938 with appendices. Bertil Sundberg, Stockholm; Marcus Stenhård, Gothenburg; Elis Larsson, Västerås and Sven Floberg, Sandviken, became members.
61. Report 8 Aug. 1939. Appendix to executive board of Metal Workers Union 10/8 1939.
62. Above report p. 1. Cf De Geer 1978, p. 153.
63. Executive board of Metal Workers Union 17/8 1939.
64. Executive board of Metal Workers Union 16/11 1939 with appendix N: report by Collin, Sundberg and Geijer 16/11 1939.
65. Decision to purchase also at executive board of Metal Workers Union 24/10 1940, 23/4 1942 and 26/3 1943.
66. Cf debate at congress of Metal Workers Union 1944, p. 227 et seq.
67. Back 1963, p. 470.
68. Back 1963, p. 471.
69. Back 1963, p. 473 et seq.
70. Policy conference on the agreement, Metal Workers Union, 1939, p. 59.
71. Policy conference on the agreement, Metal Workers Union, 1940, p. 32: Axel Åberg.
72. 29/12 1940, 15/1 1942, 19/12 1942, 26/1 1944.
73. Cf Casparsson 1951, p. 493 et seq.
74. Policy conference on the agreement, Metal Workers Union, 1942, p. 51 et seq.
75. Policy conference on the agreement, Metal Workers Union, 1942.
76. Job study committee of the Metal Trades Employers Assn. 10/9 1941. Appendix 3.
77. Working party the General council of Meta Trades Employers Assn. 29/1 1941.
78. General council of Metal Trades Employers Assn. 28/2 1941.
79. Job study committee of Metal Trades Employers Assn. 10/9 1941, Appendix 2.
80. General council of Metal Trades Employers Assn. 20/5 1941.
81. General council of Metal Trades Employers Assn. 20/5 1941.
82. Apprentices committee of Metal Trades Employers Assn. 30/5 1941, Appendix 4.
83. Apprentices committee of Metal Trades Employers Assn. 30/5 1941.
84. General council of Metal Trades Employers Assn. 11/9 et seq.
85. Job study committee of Metal Trades Employers Assn. 10/9 1941, Appendix 1.
86. Job study committee of Metal Trades Employers Assn. 10/9 1941.
87. Job study committee of Metal Trades Employers Assn. 12/12 1941.

88. Reports on conferences and the planning of conferences appear regularly in the minutes of the job study committee for the years concerned; the presentation is based on an overall impression from paragraphs in the minutes of: 31/5 1941, 9/10 1941, 12/12 1941, 18/3 1942, 28/10 1942, 10/3 1943, 26/10 1943, 11/9 1944, 28/8 1946, 20/5 1947.
89. Job study committee of Metal Trades Employers Assn. 10/9 1941, Appendix 3.
90. Job study committee of Metal Trades Employers Assn. 12/12 1941, 18/3 1942, 28/10 1942, 10/9 1943, 26/10 1943, 11/9 1944, 28/8 1946, 20/5 1947.
91. Job study committee of Metal Trades Employers Assn. 28/10 1942.
92. Job study committee of Metal Trades Employers Assn. 17/12 1947.
93. Job study committee of Metal Trades Employers Assn. 11/9 1944.
94. Job study committee of Metal Trades Employers Assn. 10/9 1941, Appendix 1.
95. Job study committee of Metal Trades Employers Assn. 10/9 1941.
96. Memorandum 27/3 1942, appendix to job study committee of Metal Trades Employers Assn. 28/10 1942.
97. Memorandum 27/3 1942.
98. Memorandum 27/3 1942.
99. Hans Thorelli—Metal Trades Employers Assn. 31/3 1942, copy attached to job study committee of this association 28/10 1942.
100. Job study committee of Metal Trades Employers Assn. 28/10 1942.
101. Ibid.
102. Ibid, appendix.
103. De Geer 1978, p. 155 et seq.
104. Job study committee of Metal Trades Employers Assn. 10/9 1943.
105. Job study committee of Metal Trades Employers Assn. 26/10 1943.

Chapter 5. Conflict and concord

1. Back 1977, p. 192.
2. Policy conference on the agreement, Metal Workers Union, 1943, p. 15.
3. Policy conference on the agreement, Metal Workers Union, 1943, p. 32.
4. Policy conference on the agreement, Metal Workers Union, 1943, p. 28.
5. Policy conference on the agreement, Metal Workers Union, 1943, p. 47.
6. Job study committee, Metal Trades Employers Assn. 26/10 1943.
7. Archives of Metal Workers Union; Negotiations, Metal Trades Employers Assn. 1943.
8. Chairman's proposal 1/12.
9. A later document from the same day, however, again limited a little the central organization's right, to the extent that it had the right to be represented at control studies only when there was a dispute that had been carried to central negotiations.
10. Report on negotiations dated 9/12 1943, archives of Metal Workers Union; Negotiations, Metal Trades Employers Assn.
11. Continued policy conference on the agreement, Metal Workers Union, Dec. 1943.
12. Continued policy conference on the agreement, Metal Workers Union, Dec. 1943, p. 73 et seq.
13. As above. Dec. 1943, p. 78; Back 1977, p. 192.
14. Proposals in archives of Metal Workers Union; Negotiations, Metal Trades Employers Assn.
15. "Memorandum on grounds for the calculation of piece wages". Appendix to job study committee of Metal Trades Employers Assn. 28/10 1942.
16. Styrman 1946, p. 249.
17. Labour Court 1944/51.
18. Ibid p. 78.
19. Ibid p. 78.
20. Ibid p. 79.
21. Ibid p. 81.
22. Some light will be thrown on the role of the piece wage question in the progress of the negotiations. For the general and dominating aspects see generally Treslow 1972 who mainly proceeds from the material of the employers' organization, and Back 1977.
23. Congress of Metal Workers Union, 1944, p. 222 et seq.
24. Motion 60 from branch in Lund.
25. Motion 61 from Huskvarna and 62 from Storfors.
26. Ibid p. 229 et seq.
27. Policy conference on the agreement, Metal Workers Union 1944, pp. 19, 21 and 121.
28. Ibid p. 118.
29. Minutes in archives of Metal Workers Union; Negotiations, Metal Trades Employers Assn. 1945.
30. Ibid.
31. Copy of letter giving notice and proposal, archives of Metal Workers Union; Negotiations, Metal Trades Employers Assn.
32. Metal Trades Employers Assn., Gen. council 28/9 1944. Cf also Back 1977, p. 234.
33. Styrman 1946, p. 255.
34. Back 1977, p. 235.
35. Back 1977, p. 259 (Thore Carlsson's statement 25–26/1), and p. 261, Oscar Westerlund 1/2.
36. Back 1977, p. 289.
37. Back 1977, p. 290.
38. The Commission's proposal 20/4 in archives of Metal Workers Union; Negotiations, Metal Trades Employers Assn.
39. Policy conference on the agreement, Metal Workers Union 1944, p. 91.
40. Back 1977, p. 290.
41. Back 1977, p. 290.
42. Back 1977, p. 294.
43. Continued policy conference, Metal Workers Union 11/5 and 18/6 1945, p. 9 et seq. and p. 86.
44. Ibid p. 86 et seq.
45. Conciliation proposal in archives of Metal Workers Union; Negotiations, Metal Trades Employers Assn. 1945.

46. Back 1977, p. 300.
47. Westerlund at policy conference on the agreement 18/6.
48. Policy conference on the agreement, Metal Workers Union 18/6, p. 38.
49. Ibid p. 41.
50. Ibid p. 54.
51. Proposal for agreement in archives of Metal Workers Union; Negotiations, Metal Trades Employers Assn. 1945.
52. Back 1977, p. 308.
53. Minutes from committee's meeting 12/3 in archives of Metal Workers Union; Negotiations, Metal Trades Employers Assn. 1946.
54. Minutes from subcommittee 26/8 1946; Negotiations, Metal Trades Employers Assn. 1946.
55. Policy conference , Metal Workers Union, 20–21/9 1946, p. 11.
56. Ibid p. 13.
57. Ibid p. 54 et seq.
58. Job study committee of Metal Trades Employers Assn. 17/12 1946.
59. See statements by Westerlund and Tyrell at continued policy conference February 1947, ibid 107 et seq.
60. Ibid.
61. Continued policy conference, Metal Workers Union 7/2 1947.
62. Ekblom's proposal 30/1 in archives of Metal Workers Union; Negotiations, Metal Trades Employers Assn. 1946.
63. Congress of Metal Workers Union, 1947. Motion 112, p. 352 et seq.
64. Westerlund's opening speech, policy conference of Metal Workers Union 1947/48, p. 3.
65. Ibid p. 101, copy of original in archives of Metal Workers Union; Negotiations, Metal Trades Employers Assn. 1947.
66. Policy conference on agreement, Metal Workers Union, 1944, p. 89.
67. Policy conference on agreement, Metal Workers Union, 1947, p. 45.
68. Ibid p. 34.
69. Appendix to technical drafting committee of Metal Trades Employers Assn. 17/12 1947. Memorandum dated 4/12.
70. Technical drafting committee of Metal Trades Employers Assn. 17/12 1947.
71. Oscar Westerlund at continued policy conference, Metal Workers Union 14/1 1948.
72. Minutes of general council, Metal Trades Employers Assn. 26–27/8 1948.
73. Back 1977, p. 428 et seq.
74. Enquiries at the Metal Trades Employers Assn. and the Metal Workers Union about documents remaining from the job study committee have produced completely negative results. Information about Ystamaskiner and Fläktfabriken is provided by a list, "The following cases have been dealt with by the job study committee" at the Metal Trades Employers Assn., Committees 69: 1.
75. The exact figures I have arrived at on a rapid examination are the following: totally 1 842 disputes, of which 172 concerned job studies in some way according to notes on the cards. 78 expressed a more general dissatisfaction; 89 concerned demands for central studies. However, I am not prepared to vouch for the exact figures without carrying out a further examination of the material. During the work I have also realized that a more ambitious treatment requires a different kind of classification. Therefore quite a new research task raises its head.

Chapter 7. The textile industry—a presentation

1. Bergsten 1957, p. 370; Engblom 1945, p. 648; Textilkursen (Textile courses) 6: 20.
2. The primary material is stored in the committee's records at the National Archives (commiittee No. 593), mainly volumes 23–26 but in certain cases also volumes 14–20.
3. See statements by Bergengren and Schmitz at the rationalization committee's conference 1936 (cf p. 00).
4. Koch 1945.
5. De Geer 1978.
6. Engblom 1945, p. 645 et seq.
7. Bergsten 1957, p. 143 et seq.
8. Olsson 1938; Tillman 1948.
9. Bergsten 1957, p. 366.

Chapter 8. The great textile strike

1. Bergsten 1957, p. 250 et seq.
2. Representative council of Textile Workers Union 17/3 1930.
3. Executive board of Textile Workers Union 16/7 1930.
4. Representative council of Textile Workers Union 22/7 1930.
5. Executive board of Textile Workers Union 7/8 1930.
6. Textile Workers Union circular No. 000.
7. Representative council of Textile Workers Union, policy conference on agreement 27/9 1943.
8. Executive board of Textile Workers Union, 24/9 1930.
9. Report in policy conference on the agreement, representative council of Textile Workers Union, 27/9 1930.
10. Re Berglund's political affiliation see Textilarbetaren No. 3, 1931, p. 49. He belonged to that Swedish communist party (the so-called Sillén communist party) which had remained as a section of the Red International after the split in the original party in 1929.
11. Representative council of Textile Workers Union 28/9 1930.
12. Textile Workers Union, circular No. 136, 5/10 1930.

13. General council of Textile Employers Assn. 15/10 1930.
14. General council of Textile Employers Assn. 24/9 1930 when the managing director reported on the district meetings and the conferences of the representatives within the Swedish Employers Confederation. With regard to Paues's reading of the situation in industry and prospects for the collective negotiations, the minutes contain only a reference to the district meetings.
15. The misgivings of the Textile Workers Union's negotiating team at the first round of negotiations 4–7 November 1930. Minutes in archives of Textile Employers Assn.; negotiation document 98/1930.
16. Cf Wennström's statement at conference of the rationalization committee 19/11 1935. See p. 242.
17. General council of Textile Employers Assn., 15/10 1930.
18. Textile Workers Union circular No. 136, 5/10 1930. Communication from Branch 36, reported to executive board 27/10 1930.
19. Executive board of Textile Workers Union 21/10 1930.
20. Minutes 2–5/11 1930 in archives of Textile Employers Assn., negotiation document No. 98/1930, copy in Textile Workers Union dossier National negotiations 1931.
21. Negotiation minutes in archives of Textile Employers Assn., negotiation document 98/1930, copy in Textile Workers Union dossier, National negotiations 1931.
22. General council of Textile Employers Assn. 18/11 1930. Changes in the draft agreement in "Memorandum on changes in the Swedish Textile Employers Association's proposal for a new collective agreement 1931, reported at meeting of 19 November 1930" in Textile Employers Association's negotiation document 98/1930.
23. "Minutes at continued negotiations ... held in Stockholm 19–22 November 1930".
24. Textile Employers Assn.–Textile Workers Union 29/9 1930. Textile Workers Union dossier, National negotiations 1931. Copy in Textile Employers Assn. negotiation document 98/1930.
25. "Minutes of negotiations before a specially appointed conciliator Allan Cederborg ... held in Stockholm 2–4 December 1930", in Textile Employers Association's document 98/1930.
26. "Memorandum of the general council's meeting 18 November 1930 concerning the question of hourly pay ...". Textile Employers Association's negotiation document 98/1930.
27. The amount is calculated in the following manner proceeding from Table 16: The piece wage excess for male workers is on average 34.3%. A cut to 15% means a reduction of approximately 20% of the hourly wage. The hourly wage in the middle schedule (of five) was 77 öre. 20% of 77 öre is about 15 öre. For the male workers 6 439 000 working hours were reported. The reduction in cost then becomes (6 439 000 ×0.15 Skr) 965 850 Skr. Calculated similarly for the female workers, where the hourly pay in the middle schedule lay at 50 öre, the reduction in cost becomes about 10 öre per working hour, or 2 538 800 Skr. In the calculations underage workers have been disregarded. As a comparison it can be noted that the total wage costs of the textile industry in 1933 are stated as 47.5 million Skr (Bergsten 1957, p. 366).
28. General council of Textile Employers Assn. 24/12 1930.
29. "Minutes of continued negotiations ... held in Stockholm 10–14 December 1930" and "Memorandum re negotiations 10–14 December ..." in negotiation document 98/1930 of Textile Employers Assn.
30. "Memorandum re negotiations 10–14 December ..." with appendices, negotiation document 98/1930 Textile Employers Assn.
31. "Alterations in the Swedish Textile Employers Association's proposal ... 12 December 1930", appendix to memorandum, see note 29 of this chapter.
32. "The workers' statement ...", appendix to memorandum, see note 29 of this chapter.
33. "The workers' reply No. 1", appendix to memorandum, see note 29 chapter.
34. "Memorandum a) with regard to the agreement with the Textile Workers Union", appendix to memorandum, see note 29 of this chapter.
35. "Reply of Textile Employers Assn. 13 December ..." appendix to memorandum, see note 29 of this chapter.
36. "The workers' reply No. 2 ..." appendix to memorandum, see note 29 of this chapter.
37. "Minutes of continued negotiations ... held in Stockholm 10–14 December 1930", Textile Employers Assn., negotiation document 98/1930.
38. General council of Textile Employers Assn. 10–14/12 1930.
39. "Report re negotiations of 10–14 December 1930", stamped 17 December 1930, appended to general committee, Textile Employers Assn. 29–30/12 1930.
40. Executive board of Textile Workers Union 18/12 1930.
41. "Minutes of continued negotiations ... held in Stockholm 29–30/12 1930", Textile Employers Assn., negotiation document 98/1930.
42. "Cederborg's memorandum No. 2, 29/12 1930", bound with answer of Textile Employers Assn. 30/12 1930, Textile Employers Assn., negotiation document 98/1930.
43. "The workers' reply to the conciliator's memorandum No. 2", bound with answer of Textile Employers Assn. 30/12 1930, Textile Employers Assn., negotiation document 98/1930.
44. "Reply of Textile Employers Assn. ... of 29 December 1930", bound with its answer of 30/12 1930, Textile Employers Assn., negotiation document 98/1930.
45. "The conciliator's request 30/12", bound with an-

swer of Textile Employers Assn. 30/12 1930, the association's negotiation document 98/1930.
46. "The Swedish Textile Employers Association's reply to the conciliator's request 30/12 1930." The association's negotiation document 98/1930.
47. "The workers' reply to the concilator's request No. 3", bound with Textile Employers Association's reply 30/12 1930, the association's negotiation document 98/1930.
48. A final offer from the conciliator 31/12 was apparently not even taken up for negotiation. (National Swedish Conciliator's Office; special appointment 1931 No. 1).
49. "Minutes of continued negotiations ... held in Stockholm 29–30 December 1930." Textile Employers Assn., negotiation document 98/1930.
50. Representative council of Textile Workers Union 1/2 1931.
51. Executive board of Textile Workers Union 2/1 1931. Textile Employers Association's circular 1/1931. The seven companies were as follows, the appropriate manager being given in parenthesis. Stockholms Bomulsspinneri & Väfveri AB, Barnängen division (Edvard Ericsson): AB Förenade Yllefabrikerna, Norrköping (Carl Wahren); Borås Wäfveri AB's factories in Borås and Skene, (Axel Bergengren, chairman of Textile Employers Assn.); Holma-Helsinglands Linspinneri & Väfveri AB's factories in Fors and Madängsholm and Almedal-Dalsjöfors AB's factories in Almedal and Dalsjöfors (Axel Leman, vice-chairman of Textile Employers Assn.); AB Sveriges Förenade Trikåfabriker's divisions A, D, E and BS in Borås and G in Mölndal and N in Malmö (Carl Eiserman Jr); Nordiska Maksinfilt AB in Halmstad (G. Westerberg).
52. Bergsten 1957, p. 263.
53. General council of Textile Employers Assn. 29–30/12 1930.
54. Circular, Textile Employers Assn. No. 25/1930, dated 31/12 1930.
55. Circular, Textile Employers Assn. No. 1/1931. Also Bergsten 1957 has misunderstood the wage cut as a countermeasure! (263). The same version is given by Olsson 1938, p. 219.
56. General council of Textile Employers Assn. 3/1 1931.
57. General council of Textile Employers Assn. 3/1 1931.
58. The decision of these unions was notified to the other side in communications of 8 and 9 January. Textile Employers Association's general council 11–13/2 1931.
59. Executive board of Textile Workers Union, 5/1 1931.
60. Circular 144, Textile Workers Union, 6/1 1931.
61. Circular, Textile Employers Assn. No. 3/1931, dated 7/1 1931.
62. Executive board of Textile Workers Union 10/1 1931, Textile Employers Association's circular No. 3/1931. Discussions with the Minister for Social Affairs took place on 7 January.
63. General council, Textile Employers Assn. 11–13/2 1931.
64. Information from Textile Employers Assn. to the National Social Welfare Board, social statistics division. Copy in Textile Employers Association's negotiation documents 98/1930, dated 20/10 1931; Olsson 1938, p. 220.
65. Textile Workers Union, circular No. 150, 18 March 1931.
66. At the middle of February K. J. Olsson again met Edvard Johansson, chairman of the Trade Union Confederation, to discover his attitude to a nationwide sympathetic action in favour of the textile workers. The regional trade union organization of Stockholm had raised the matter with Olsson and said it was prepared to take the initiative in this direction. But Edvard Johansson had remained comparatively indifferent and thought it would be unwise to start fund collections whilst negotiations were in progress although it might be suitable for the representative council of the Trade Union Confederation to be summoned to consider the question of a financial contribution should the negotiations before the conciliation commission break down. (Executive board of Textile Workers Union 20/2 1931).
67. Olsson 1938, p. 220 et seq.
68. Information to social statistics; cf. note 64.
69. Executive board of the Textile Workers Union, 26/1 1931.
70. Official statistics of Sweden, Stoppages and Collective Agreements 1931, p. 130.
71. Convocation notice 7/2 1931 in Textile Workers Union's dossier, National agreement negotiations 1931. No documents are on file with either party on the discussions from 12/2 to 8/3.
72. Pencilled minutes from negotiations at National Conciliators Office: special appointment 1931 No. 1.
73. General council, Textile Employers Assn. 11–13/2 1931. Executive board, Textile Workers Union 20/2 1931.
74. See note 2.
75. Executive board, Textile Workers Union 20/2 1931.
76. Textile Workers Union, dossier National agreement negotiations 1931, draft of request.
77. "View of the workers' side on the conciliator's proposal of 2 March 1931" dated 4/3 1931. National Conciliators Office, special appointment 1931 No. 1.
78. Textile Workers Union, dossier National agreement negotiations 1931, dated 6/3 1931 (also in National conciliators Office, special appointment 1931 No. 1).
79. Handwritten draft of minutes of negotiations 5/3. National Conciliators Office, special appointment 1931 No. 1.
80. Request of the conciliation commission 8 March—also printed version "The conciliation commission's proposal for settlement ..." Textile Workers Union, dossier National agreement negotiations 1931.
81. General council, Textile Employers Assn. 9/3 1931.

82. "Memorandum on the general council's treatment ...", 10/3 1931, Textile Employers Assn., negotiation document No. 98/1930.
83. "To the conciliation commission in the textile conflict ...", dated Stockholm 16 March 1931, Sweden's Textile Employers Assn., negotiation document No. 98/1930, Textile Employers Assn.
84. Executive board, Textile Workers Union 9/3 1931.
85. Representative council, Textile Workers Union 10/3 1931.
86. Textile Workers Union, circular No. 149 dated Norrköping 10 March 1931.
87. 6926 members voted for approval of the proposal and 11989 against; Olsson 1938, p. 222.
88. Executive board, Textile Workers Union 13/3 1931.
89. In the Textile Workers Union dossier, national agreement negotiations 1931, there is a memorandum with 10 remarks specified and headed "Comments on the conciliation board's proposal ...". Of these 10, the two mentioned have been made the main points in the union's reply to the conciliation commission. See transcript (copy) in the Textile Employers Assn. document No. 98/1930.
90. Textile Workers Union, representative council, 16/3 1931.
91. Private letter at National Conciliators Office, special appointment 1931 No. 1.
92. On the letter Cederborg has noted "Yes!" or "No!".
93. Minutes at National Conciliators Office, special appointment 1931 No. 1.
94. Minutes at National Conciliators Office, special appointment 1931 No. 1.
95. See note 1.
96. "Minutes of continued negotiations ... held in Stockholm 19 and 20 March 1931", Textile Employers Assn., document 98/1930.
97. Bergsten 1957, p. 268.
98. Bergsten (p. oo) states that the disturbances in connection with the strike emanated from syndicalist circles in Borås. However, this is not reflected in the central material at the Textile Employers Assn., which has been utilized for the present investigation.
99. See Hartmut Apitzsch "RFO-Red Union Opposition" in Arkiv No. 2, 1972.
100. Textile Workers Union, circular 147.
101. Leaflet "Kamp mot nederlagsorganisatörerna" (The struggle against the organizers of defeat!). Textile Employers Assn., document No. 98/1930.
102. The demands on which the communist party carried on its fight against the buyers of labour and against the social democrats in the reformistic trade union management were the following:

"1) 25% increase in wages.

2) A further 50% bonus for workers in trades injurious to health.

3) 25% bonus for two shift work.

4) 50% bonus for night shift.

5) Equal pay for equal work regardless of sex, age and other differences.

6) 14 days holiday during July or August ... with full pay.

7) A 7-hour working day with no reduction in pay.

8) A 6-hour working day for young people under 18 years of age and for work specially injurious to health.

9) Abolition of piecework.

10) No reduction in pay at stoppage because of breakdown in machinery or bad planning of management and similar causes.

11) In the curtailing of operations there may be no sacking or laying-off of individual workers but available work shall be distributed equally and wages be paid for full time.

12) Rest period in the work when the work is hectic.

13) Full guarantee of reemployment after compulsory national service, illness or similar occasions.

14) One month's leave before and after delivery with retained pay for pregnant women.

103. Textile Workers Union, circular No. 151, 21 March 1931.

Chapter 9. From 1931 to 1936

1. Representative council, Textile Workers Union 20/5 1932.
2. Representative council, Textile Workers Union 27/6 1933.
3. Executive board, Textile Workers Union 26/9 1933.
4. Representative council, Textile Workers Union 26/5 1934.
5. Representative council, Textile Workers Union 24–25/5, 2–3/9 and 25/9 1935.
6. General council, Textile Employers Assn. 22/8 1931. The general council also decided to approach the different districts of the association to obtain their views on the question. At the meeting in mid-September the managing director, Erik Paues, was able to report that all districts shared the view of the situation held by the general council. The decision now lay with the general council and the meeting of the association.
7. General council, Textile Employers Assn. 14/9 1931.
8. General council, Textile Employers Assn. 25/9 1931.
9. General council, Textile Employers Assn. 25/9 1931 with appendices.
10. General council, Textile Employers Assn. 30/9 1931.
11. General council, Textile Employers Assn. 30/9 1931.
12. General council, Textile Employers Assn. 3/9 1932.
13. Cf Official Statistics of Sweden, Stoppages and Collective Agreements 1932 and 1933.
14. General council, Textile Employers Assn. 14/9 1933.

15. General council, Textile Employers Assn. 26/9 1934.
16. General council, Textile Employers Assn. 26/9 1934.
17. General council, Textile Employers Assn. 26/9 1934.
18. Conference on the textile industry held by the Rationalization Committee, cf. p. 242 and notes to that chapter.
19. Textile Employers Assn. Appendix to general council 26/9 1934.
20. Representative council, Textile Workers Union, 24–25/5 1935.
21. "Agreement ..." etc., Stockholm and Norrköping 11 June 1935. Textile Employers Assn., document No. 85/1935.
22. Representative council, Textile Workers Union, 2–3/9 1935.
23. General council, Textile Employers Assn., 18/9 1935.
24. Minutes of meeting ... held in Stockholm, Tuesday 17 and Wednesday 18 September 1935. Textile Employers Assn., document No. 85/1935.
25. Minutes ... 17 and 18 September 1935. Textile Employers Assn., document 85/1935.
26. Representative council, Textile Workers Union 25/9 1935.
27. General council, Textile Employers Assn. 18/9 1935. The only positive demand the association was to put forward in the negotiations dealt with the placing of Furulund to a lower schedule.
28. Cf "the stretcher conflict" 1934. See Textile Workers Union, annual report 1934, p. 74 et seq.
29. "Minutes ... held in Stockholm on Thursday 26 and Friday 27 September 1935. Textile Employers Assn., document 85/1935.
30. "Agreement" dated 27 September 1935. Textile Employers Assn., document 85/1935.

Chapter 10. Disputes about the Bedaux system and job studies 1947

1. The Bedaux system was named after Charles Eugene Bedaux (1886–1944). Born in France he emigrated to the USA where he became naturalized. He organized and led firms of consultants in company management, when he developed a method comprising a system for the calculation of performance and the earnings connected with it. In 1937 Bedaux wound up his activities in the USA and returned to France. During the second world war he cooperated with the Vichy regime and was arrested by the allied troops in North Africa and tried in the USA for treason. He committed suicide. The basis of Bedaux's system is a carefully calculated performance unit called a Bedaux unit (B), which corresponds to what an average worker could deal with in a minute, including time for rest. In order to decide the actual content of a Bedaux unit for different work, for example how many pieces are produced, detailed job studies combined with a subjective estimate of the working speed during measurement. The normal performance thus is that a worker carries out 60B/hour. In accordance with the method it should however be possible for the workers to achieve 80B/hour, when the excess is awarded a bonus according to a fixed scale. So the Bedaux system was really a kind of bonus piece wage. (Re Bedaux himself: Encyclopaedia Britannica, Volume 3, 1964, p. 360d. Re the Bedaux system: Företagsekonomisk Handbok, Part II, 1945, p. 358 et seq).
2. Textile Workers Union. Annual report 1934 p. 89.
3. Minutes of negotiations between the Swedish Textile Employers Assn. and the Swedish Textile Workers Union concerning the application of the so-called Bedaux system at Malmö Strumpfabrik, held at the offices of the company in Malmö Thursday 3 May 1934. Textile Employers Assn. document 65/1933; general council, Textile Employers Assn. 18/4 1934; Textile Workers Union, annual report 1934, p. 89 et seq.
4. It is noted that the minutes of the Textile Workers Union usually entitle the dispute "job studies".
5. "To the personnel at Rydboholms Aktiebolags-fabriker", a stencilled notice signed "Rydboholm 8 August 1933 Claes Eriksson", Textile Employer Assn., document 64/1933.
6. Rydboholms Aktiebolag–Textile Employers Assn. 6/9 1933, according to the minutes of the negotiations 25 October 1933, paragraph 2, Textile Employers Assn., document 64/1933.
7. Executive board, Textile Workers Union 14/9 1933.
8. Executive board, Textile Workers Union 14/9 1933.
9. Rydboholms AB–Textile Employers Assn., 6/10 1933. Copy in minutes of negotiations 25/10 1933.
10. Textile Employers Assn–Textile Workers Union 9/10 1933. Copy in minutes of negotiations 25/10 1933. Textile Employers Assn., document 64/1933.
11. Textile Workers Union–Textile Employers Assn. 11/10 1933. Copy in minutes of negotiations 25/10 1933. Textile Employers Assn., document 64/1933.
12. Textile Workers Union, executive board 18/10 1933.
13. Minutes of negotiations 25/10 1933. Textile employers Assn., document 64/1933.
14. The minutes of the negotiations 25/10 render Paues' point of view in the following way: "In the settlement of piece wage prices to confirm such grounds as the Bedaux system has calculated with has not of itself meant any deviation from the provisions of the agreement, provided that the piece wage was fixed. But should it again be a question of a variable piece wage, for which the hourly pay is to constitute a guarantee, there has not been scope for such a thing in the agreement. If the question again refers to work paid by the hour, then according to the Bedaux system even this became comparable to a variable piece wage, and this must be considered to be contrary to the agreement, which presumes that the worker for fixed hourly pay does so much and so good a work as he is capable of, independent of possible bonuses."

15. The position was put to the executive board of the Textile Workers Union at the meeting 17/11. The representative reported that he had offered the local branch of the union two proposals for a decision. The branch had selected the second of these, which meant that the workers were not willing to cooperate in job studies as they were now carried out but were prepared to discuss some simpler form of job study with the management of the factory.
16. Branch 65—Rydboholms Aktiebolag 29/11 1933, added to the negotiations minutes 16/1 1934. Textile Employers Assn., document 64/1933. As appears, the contents of the addition are somewhat different from that stated by the Textile Employers Association's general council 17/11 1933. See previous note.
17. Rydboholms Aktiebolag (Claes Eriksson)—Branch 65 Textile Workers Union 9/12 1933, copy in negotiation minutes 16/1 1934. Textile Employers Assn., document 64/1933.
18. Branch 65 Textile Workers Union (chairman Nils Erikson)—Rydboholm Aktiebolag 27 December 1933. Copy in minutes of negotiations 16/1 1934. Textile Employers Assn., document 64/1933. The executive board had also made the branch a proposal in this communication. Cf Textile Workers Union, executive board 27/12 1933.
19. Minutes of negotiations between the Swedish Textile Employers Assn., and the Swedish Textile Workers Union regarding the refusal of the workers at Rydboholm Aktiebolag to provide certain of the informations requested by the factory management concerning production, held in Stockholm Tuesday 16 January 1934. Textile Employers Assn., document 64/1933.
20. Executive board, Textile Workers Union 21/1 1934.
21. Textile Employers Assn.,–Labour Court 1/3 1934, appendix C, copy in Textile Employers Assn. document 64/1933.
22. Textile Employers Assn.,–Labour Court 1/3 1934, copy in Textile Employers Assn. document 64/1933.
23. Textile Workers Union–Labour Court 9/3 1934. Signed copy in Textile Employers Assn. document 64/1933.
24. Rydboholms Aktiebolag–Textile Employers Assn., 15/3 1934. Textile Employers Assn. document 64/1933.
25. Copy of Labour Court's judgement 38/1934. Textile Employers Assn. document 64/1933.
26. "Extract from minutes written at the Labour Court's meeting ... 17 March 1934." Copy in Textile Employers Assn. document 64/1933.
27. Annual report Textile Workers Union 1935, p. 76 et seq.
28. There was a dispute at Gårda Fabrikers AB about, among other things, a payment based on the Bedaux system. At the main negotiations it was decided to abandon this system at the factory (Textile Workers Union, annual report 1939, p. 125).
29. Minutes of negotiations between the Swedish Textile Employers Assn. and the Swedish Textile Workers Union concerning job study activity at Malmö Yllefabriks AB, held at Malmö 30 December 1947. Textile Employers Assn. document 55/1947.

Chapter 11. Collective negotiations 1936–1938

1. Textile Employers Assn., general council.
2. General council, Textile Employers Assn., 24/9 1936. The technical conference met 14–15/9 1936.
3. Re technical conferences see Bergsten 1957.
4. Minutes of the first meeting are missing from the series of minutes of the technical conferences. The presentation here is based on references in the minutes of the general council 24/9 1936.
5. General council, Textile Employers Assn., 24/9 1936. Written reply to Textile Workers Union, dated 30/9 1936 in Textile Workers Union, dossier National agreement negotiations 1937.
6. Textile Workers Union, representative council 17–18/7, 5/9 1936.
7. Re changes, see Olsson 1938 p. 234 et seq.
8. Textile Workers Union, general council 12/9 1936.
9. Textile Workers Union, executive board 22/9 1936.
10. Textile Workers Union's proposal for collective agreement etc., received at Textile Employers Assn. 30 September 1936. The association's document 71/1936.
11. The survey of the employers' reactions is based on the notes made by Erik Paues on the filed copy of the Textile Workers Union's draft agreement.
12. Re request of the conciliator, Textile Employers Assn. and Textile Workers Union–National Board of Social Welfare 15/10 1936; re start of negotiations, Ekblom–Textile Employers Assn. 19/10 1936, Textile Employers Assn. document 71/1936.
13. Textile Employers Assn. 24/10 1936.
14. Paues' notes, see Note 11.
15. Textile Employers Assn., general council 24/10 1936; the association's draft agreement is attached to the general council's series of minutes for 1936.
16. Textile Workers Union, executive board 11/11 1936. Textile Employers Assn., general council 11/11 1936.
17. Textile Workers Union, executive board 11/11 1936.
18. Textile Workers Union, executive board 1/12 1936. As the account of this phase in the negotiations is based on the report of the workers' negotiation tema the arguments of this side are shown in more detail. Negotiations took place on 20–25/11 1936. Cf also Textile Employers Assn., general council 18–21, 23–25/11 1936.
19. Textile Employers Assn., general council 3–5, 7–8/12 1936.
20. Stencilled proposal entitled Textilindustrien (The Textile Industry), dated 8 December 1936. Textile Workers Union, dossier National agreement negotiations 1937.

21. Stencilled proposal entitled Textilindustrien (The textile industry), dated 8 December 1936, p. 3. Textile Workers Union, dossier Nationa agreement negotiations 1937.
22. Cf Bergsten 1957 p. 372.
23. Textile Workers Union, general council 10/12 1936 p. 55.
24. Textile Workers Union, general council 10/12 1936 p. 42 (Wennström). In the conciliator's report for 1936 the wage question is stated improperly as the cause of the breakdown. (Official Statistics of Sweden, Stoppages and Collective Agreements 1936 p. 138.)
25. Textile Workers Union, general council 10/12 1936 p. 48 (Wennström).
26. Textile Workers Union, general council 10/12 1936 p. 44 (Wennström).
27. Textile Workers Union, general council 10/12 1936 p. 38.
28. Textile Workers Union, general council 10/12 1936 p. 51.
29. Textile Workers Union, general council 10/12 1936, p. 53.
30. Textile Workers Union, general council 10/12 1936 p. 46.
31. Textile Workers Union, general council 10/12 1936 p. 53.
32. Official Statistics of Sweden, Arbetsinställelse och Kollektivavtal (Stoppages and Collective Agreements) 1936, p. 138. Following the negotiations in December, Ekblom was not exactly persona grata in the workers' camp, as is shown by Wennström's speech at the general council's meeting 12/1 1937.
33. Textile Employers Assn., general council 7–11/1 and 22–24/1 1937.
34. The proposal is located in Textile Employers Assn., general council, appendix 1 of the minutes 7–11/1 1937 and also in Textile Workers Union, general council 12/1 1937.
35. Textile Employers Assn., general council 7–11/1 1937.
36. The workers' answer is located in Textile Employers Assn., general council, appendix 2 of minutes 7–11/1 1937. Wennström's speech in Textile Workers Union, general council 12/1 1937.
37. The Trade Union Confederation's Press Information No. 4, 1937, Stockholm 12 January 1937. Stencilled copies in Textile Workers Union, dossier National agreement negotiations 1937. Re genesis see Textile Workers Union, general council, 12/1 1937, p. 1, Wennström's report.
38. Textile Workers Union, general council 12/1 1937 p. 6.
39. Textile Workers Union, general council 12/1, continued discussions 13/1.
40. Cf Bergsten 1957, p. 211 et seq.
41. Textile Workers Union, general council 12/1 1937.
42. Textile Employers Assn., general council 22–24/1 1937. An illustrative discussion of the principles behind the selection of factories was held at the meeting of the Textile Workers Union's general council on 12/1 1937. Bergengren's factories were to be brought out on strike as he was influential in a number of industrial organizations. A spread across the various branches and across the country was desirable as only certain factories were to be brought out, so that in this way the competition between the companies could be used to test loyalty to the organization. Even such criteria appeared as managements' good will towards the workers.
43. Textile Employers Assn., document 71/1937.
44. Textile Workers Union, general council 27–28/5 1938.
45. Textile Workers Union, general council 1/9 1938.
46. Textile Employers Assn., general council 5–7/9 1938.
47. Textile Workers Union, general council 1/9 1938 and Textile Employers Assn., general council 5–7/9 1938.
48. 5–7 and 13–15/9. Textile Employers Assn., general council, same days.
49. Textile Employers Assn., general council 5–7/9 1938.
50. Textile Employers Assn., general council, 13–15/9 1938, appendix 15; Textile Workers Union, general council 24/9 1938, Wennström's report.
51. Textile Workers Union, general council 24/9 approved the proposal unanimously, as the policy conference on the agreement the next day (the minutes are missing!). The employers established the proposal firmly in the districts and the general council decided to subscribe to the agreement 26/9 1938.
52. Textile Employers Assn., general council 22/3 1938.

Chapter 12. Collective negotiations in the 1940s

1. The workers' draft agreement attached to minutes of the negotiations 7/11 etc. 1939. Textile Employers Assn., document 69/1939.
2. Minutes of negotiations 11/1 1940 with attached preliminary agreement 30/12 1939. Textile Employers Assn., document 69/1939.
3. Agreement 29/1 1941. Textile Employers Assn., document 42/1940.
4. Minutes of negotiations 29/1 1941. Textile Employers Assn., document 42/1940.
5. Minutes of negotiations 20–21/11 1941. Textile Employers Assn., document 37/1941.
6. Agreement 30/1 1942. Textile Employers Assn., document 37/1941.
7. Minutes of negotiations 19–20/11 and 4–5/12 1942. Textile Employers Assn., document 45/1942.
8. Textile Employers Assn., general council, 4–5/12 1942.
9. Agreement 16/12 1942. Textile Employers Assn., document 45/1942.
10. Agreement—see Textile Employers Assn., general council 15–16/1 1944. Bergsten 1957, p. 306.

11. Minutes of negotiations 13–14/12 1944. Textile Employers Assn., document 92/1944.
12. Agreement 3/12 1945. Textile Employers Assn., document 92/1944.
13. Here the account is based on Bergsten 1957, p. 316, as complete minutes of the preliminary negotiations in November 1945 are missing at the Textile Employers Assn.
14. Bergsten 1957 mistakenly gives this model as a new one for the agreement concerned, p. 316.
15. Bergsten 1957, p. 372 et seq.
16. The following account also is based mainly on Bergsten 1957, p. 319 et seq.

Chapter 13. Independent arguments of the parties

1. Textilarbetaren No. 1, 1934, p. 8.
2. For the motion and its treatment in the Riksdag see De Geer 1978, p. 265 et seq.
3. Textilarbetaren No. 2, 1934, p. 41 et seq.
5. About these things, see De Geer 1978, passim.
6. Textilarbetaren No. 2, 1935, p. 58.
7. Gillis Elfstrand was corresponding secretary for Branch 99 during 1933–35. In all probability he was therefore also its chairman. He is stated to have been "a stable social democrat" and worked at Uddebo Möbeltyg Väfveri AB, where it is likely that there had been no job studies until the middle of the 1930s. Information from the earlier branch cashier, Gustav Larsson, through the agency of the earlier branch secretary, Georg Lindgren of Norrköping. During the period with which we are concerned here Larsson was leader of district studies in the Social Democratic Youth of Sweden (SSU), southern Älvsborg and later chairman of the Textile Workers Union branch at Kinna. He was an official of the union from 1950.
8. Congress of Textile Workers Union 1936. Motion 53.
9. Congress of Textile Workers Union 1936, p. 256 et seq.
10. Minutes in the rationalization committee's archives, National Archives, committee No. 593. The conference was held 19/11 1936.
11. Minutes in the rationalization committee's archives, National Archives, committee No. 593.
12. Bergsten 1957.
13. The statement is written by Kaj Åberg, but a number of ideas and phrases are recognized from Engblom 1945.
14. Regarding the attitude to courses at the Institute for the Training of Foremen (ALI) see Textile Employers Assn., general council 22/9 1943, where it was reported that in collaboration between the association and ALI a special time and motion study course was being planned for the textile industry on 25/10–20/11 1943 with about 15–20 places. ALI was stated to be prepared to repeat the course later. A number of companies in Western Sweden had said they were prepared to accept study visits. The links between the textile industry and ALI were also older. Professor Rudolf Anderberg was appointed director of ALI's psychotechnical activities from 1937 for investigations in the textile industry. (Textile Employers Assn., general council 19/4 1939; De Geer 1978, p. 145.)
15. Textile Employers Assn.—Swedish Employers Confederation 3/6 1946, approved by Textile Employers Assn., general council, and attached to the minutes 26–27/9 1946.
16. Textile Employers Assn., general council. 26–27/9 1946.
17. Textile Employers Assn., general council. 13/11 1947.

Chapter 15. Path to the job study agreement 1948

1. Schiller 1974, p. 65.
2. Motion in first chamber No. 211, second chamber No. 47.
3. Ibid p. 3.
4. Ibid p. 5.
5. Motion second chamber No. 186.
6. In the second chamber's committee some of the conservative members made reservations with reference to the cooperation already started between the Employers Confederation, the Federation of Swedish Industries and the Trade Union Confederation.
7. Committee report 12, compilation (samling) 1, section 15, and report 13, compilation 2, section 5.
8. Second chamber 4/5 1945, minute 19: 93; first chamber 2/6 page in minute 25: 55.
9. Ibid p. 94 et seq.
10. Ibid p. 100.
11. Riksdag communication No. 509.
12. See De Geer 1976.
13. Casparsson 1951.
14. At the time the following were members of the committee: Swedish Employers Confederation: Gustaf Söderlund, C. J. Malmros, Fritiof Söderbäck, Ivar O. Larsson, Nils Danielsen and Elof Ericsson. Swedish Trade Union Confederation: August Lindberg, Gunnar Andersson, Oskar Karlén, Hilding Molander, Oscar Westerlund, Gustav Vahlberg and Ernst Falk. The secretary was Alvar Odholm, Employers Confederation, and Arnold Sölvén, Trade Union Confederation. Hjalmar Åselius, Gösta Ekelöf, Gösta Ingelman and Erik Collin were present in addition when the job study question was discussed.
15. Regarding joint works council's agreement see Casparsson 1951, p. 429 et seq; Schiller 1974, p. 65 et seq.
16. Cf Casparsson 1951, p. 346 et seq.
17. Report of Ekelöf's speech, appendix to Labour Market Committee 31/8 1945.
18. Regarding ALI's position see De Geer 1978, p. 153 et seq.
19. Labour Market Committee minutes (AMK) 31/8 1945.
20. Labour Market Committee's minutes 22–23/11 1945. Delegated by employers Bertil Kugelberg, Nils Danielsen and Gösta Ingelman, from Trade Union Confederation, Gustav Vahlberg, Erik Collin and Ture Flyboo.

21. Sölvén: Memorandum for Labour Market Committee's treatment of question about industrial democracy.
22. Cf especially p. 6.
23. Cf especially p. 6: the workers' pride.
24. Ibid—8.
25. Ibid p. 9 et seq.
26. Minutes of Labour Market Committee 25–26 June 1946.
27. Appendix to minutes of Labour Market Committee 25–26 June 1946.
28. Ibid p. 2 et seq.
29. Minutes of Labour Market Committee 15–16 April 1947.
30. Ibid.
31. Labour Market Committee 4–5 November 1947.
32. Labour Market Committee 25–26 May 1948.
33. Labour Market Committee 26/8 1948. The complete agreement is given as appendix 3.

Chapter 16. In retrospect

1. De Geer 1978, passim.
2. De Geer 1978, passim.
3. Styrman 1946.
4. Edlund 1974, Eklund 1974.
5. Edlund 1974.

References

1. Unpublished source material

Swedish National Archives (Riksarkivet):
National Conciliators Office, Special commissions
Rationalization committee's archives, committee 593

The Folk Movements' archives in Norrköping (Folkrörelsearkivet):
Archives of the Swedish Textile Workers Union:
minutes of representative councils
minutes of general council
series of negotiations on National agreements
circulars

Swedish Trade Union Confederation's archives, Stockholm (LO):
Labour Market Committee's minutes and documents

Swedish Metal Workers Union's archives, Stockholm (Svenska Metallindustriarbetarförbundet):
minutes of executive board and general council
series of negotiations with Swedish Metal Trades Employers Association

Swedish Textile Employers Association, Stockholm (Sveriges Textilindustriförbund):
minutes of the board
negotiation documents

Swedish Metal Employers Association, Stockholm (Sveriges Verkstadsförening):
minutes of the general council
apprentices committee/technical drafting minutes
minutes of the job study committee
conflicts register
list of committees

Wäfveriaktiebolaget, Borås:
Stencilled history: Koch, Gunnar, job studies at Borås Wäfveri AB, 1945

2. Published source material

Contemporary publications, revised versions, reference books, periodicals and papers

Apitzsch, Hartmut RFO – Röd facklig opposition (Red union opposition) in *Arckiv* No. 2, 1972

Attityder till tekniken (Attitudes to technology) Bank of Sweden Tercentenary Foundation 1978: 6, Stockholm 1978

Back, Per Erik Svenska metallindustriarbetareförbundets historia 1925–1940 (History of Swedish Metal Workers Union 1925–1940) Vol. III, Stockholm 1963

Back, Per Erik Svenska metallindustriarbetareförbundets historia 1940–1956 (History of Swedish Metal Workers Union 1940–1956) Vol. IV, Stockholm 1977

Bergsten, Nils Sveriges Textilindustriförbund (Swedish Textile Employers Association), Stockholm 1957

Casparsson, Ragnar LO under fem årtionden II 1924–1948 (The Swedish Trade Union Confederation over five decades II 1924–1948), Stockholm 1951

Casparsson, Ragnar LO – Bakgrund, utveckling och verksamhet (The Swedish Trade Union Confederation – Background, development, activities), Halmstad 1966

Dahlström, Edmund Information på arbetsplatsen. Sociologiska studier av företagets kommunikationsproblem (Information at the workplace. Sociological studies of the company's communication problem), Stockholm 1956

De Geer, Hans Arbetsfreden i Sigtuna (The industrial peace discussions in Sigtuna) in *Scandia* 1976

De Geer, Hans Rationaliseringsrörelsen i Sverige (The rationalization movement in Sweden), Uddevalla 1978

Edlund, Sten Perspektiv på arbetsdomstolen (The Labour Court—a perspective), in *Edlund–Eklund* Rätt och arbetsgivarmakt. Två studier i rättsprocessen på arbetsmarknaden (Law and employer power. Two studies on the process of law on the labour market), Stockholm 1974

Eklund, Per Rätten i klasskampen (Law in the class struggle), Stockholm 1974

Encyclopaedia Brittanica

Engblom, Alex Arbetsstudier inom textilindustrin (Work studies in the textile industry) in *Företagsekonomisk handbok* (Handbook of Business Economics), 1945

Fackföreningsrörelsen och den tekniska utvecklingen. Rapport till 1966 års LO-kongress (The trade union movement and technical development. Report to 1966 congress of the Swedish Trade Union Confederation), Stockholm 1966

Företagsekonomisk handbok del II (Handbook of Business Economics part II), published by C. A. Strömberg, Stockholm 1945

Gardell, Bertil Arbetsinnehåll och livskvalitet (The content of work and life quality), Stockholm 1976

Gardell, Bertil Produktionsteknik och arbetsvillkor (Production engineering and conditions of work). In *Attityder till tekniken* (Attitudes to technology), Bank of Sweden Tercentenary Foundation 1978: 6, Stockholm 1978

Geijer, Lennart–Schmidt, Folke Arbetsgivare och fackföreningsledare i domarsäte (Employers and trade union leaders in the judge's seat), Lund 1958

Handbok i industriell driftsekonomi och organisation (Handbook of business economics and organization) ed. Tarras Sällfors, Stockholm 1945

Kennerström, Bernt Mellan två internationaler (Between two internationals) Kristianstad, Sweden 1974

Korpi, Walter Arbetarklassen i välfärdskapitalismen (The working class in welfare capitalism), Kristianstad, Sweden 1978

Lindgren, John Svenska metallindustriarbetareförbundets historia 1888–1905 (History of the Swedish Metal Workers Union, 1888–1905) Vol I, Stockholm 1938

Lindgren, John–Tingsten, Herbert–Westerståhl, Jörgen Svenska Metallindustriarbetareförbundets historia 1906–1925 (History of the Swedish Metal Workers Union 1906–1925) Vol. II, Stockholm 1948

Meidner, Rudolf Samordning och solidarisk lönepolitik (Coordination and unanimous pay policy), Stockholm 1974

Olsson, K. J. Textilarna (The Textile Workers), Norrköping 1938

Olsson, Sven Järnbruksförbundet 1906–1956 (The Swedish Iron and Steel Works Association 1906–1956), Stockholm 1958

Olsson, Ulf Lönepolitik och lönestruktur (Pay policy and pay structure), Gothenburg 1970

Riksdagens protokoll med bihang (Official reports of the proceedings and debates of the Swedish Riksdag with annexes)

Schiller, Berndt LO, § 32 och företagsdemokratin (Swedish Trade Union Confederation, paragraph 32 and industrial democracy) Stockholm 1974

Segerstedt, Torgny–Lundquist, Agne Människan i industrisamhället (Man in industrial society), Stockholm 1952

Sociological research

Styrman, Georg Verkstadsföreningen (The Swedish Metal Trades Employers Association) 2nd edition, Stockholm 1946

Svenska Metallindustriarbetareförbundet (The Swedish Metal Workers Union) Avtalskonferenser för verkstadsindustrin; Verkstadsarbetarkonferenser; Verkstadskonferenser 1922–1948. (Policy conferences on the agreement for the metal trades industry. 1922–1948). Kongressprotokoll 1922–1948 (Congress minutes 1922–1948)

Svenska Teknologföreningen (The Swedish Association of Engineers and Architects) 1861–1936. Biographies. Stockholm 1937

Svenska Textilarbetareförbundet i samarbete med *Brevskolan:* Textilkursen (The Swedish Textile Workers Union in collaboration with Brevskolan: The textile course) Stockholm 1946–1949

Svenska Textilarbetareförbundet (The Swedish Textile Workers Union) Annual Reports 1926–1948, Norrköping, Sweden, each year

Textilarbetaren (The Textile Worker), periodical

Tillman, Bo Textilarna II 1938–1948 (The textile workers II 1938–1948), Norrköping, Sweden 1948

Treslow, Kjell Verkstadsstrejken 1945. En studie av arbetsmarknadens förhandlingssystem vid konflikt. (The Metal Trades' conflict of 1945. A study of the collective bargaining system in a labour market conflict.) Stencilled, Institute for political science, University of Stockholm. Stockholm 1972

Victorin, Anders Lönenormering genom kollektivavtal (Standardization of wages through collective agreements), Stockholm 1973

List of tables

List of diagrammes

Index of persons

(This index does not include authors only referred to in the notes)